Low Temperature Behaviour of Solids: An Introduction

Solid-State Physics

Edited by **L. JACOB,** *D.Sc., Ph.D., F.Inst.P.*
Senior Lecturer in Natural Philosophy,
University of Strathclyde

Electron and Ion Emission from Solids
R. O. Jenkins and W. G. Trodden

The Thermal Properties of Solids
N. J. Goldsmid

Solid Semiconductors
A. K. Jonscher

Electrical Conduction in Solids
H. Inokuchi

Low Temperature Behaviour of Solids
R. G. Scurlock

Low Temperature Behaviour of Solids: An Introduction

R. G. SCURLOCK

M.A., D. Phil.,
Lecturer in Physics
University of Southampton

LONDON: Routledge and Kegan Paul Ltd.
NEW YORK: Dover Publications Inc.

First published 1966
in Great Britain by
Routledge and Kegan Paul Ltd.
and in the USA by
Dover Publications Inc.
180 Varick Street
New York 10014

Printed in Great Britain
by W. & G. Baird Ltd., Belfast.

Contents

Figures

Tables

Preface

This book provides an elementary introduction to the behaviour of solids, at temperatures ranging down from room temperature. It is aimed at the level of the second or third year undergraduate student in science and engineering, and provides a concise account of some of the more important properties of the solid state. A strict mathematical approach is avoided, and discussion is limited to qualitative, order of magnitude, explanations of low temperature behaviour.

The reader is assumed to have an elementary acquaintance with vectors, and with the principles of thermodynamics. References are given at the end, for further reading.

After introducing the temperature spectrum, the various types of specific heat phenomena are described. A qualitative treatment of lattice vibrations and electrons in metals is made at some length to help understand lattice and electronic specific heats, and transport phenomena described in the following chapter. An outline of phonon and electron heat conduction is described, followed by a section on electrical conduction in metals.

The chapter on superconductivity gives an introduction to Types I and II superconductors, high-field superconductors and 'super-magnets', and the superconducting energy gap. The next chapter ranges over the wide temperature spectrum covered by magnetic phenomena, and includes an account of electron and nuclear paramagnetism, adiabatic demagnetisation of electron and nuclear spins, and the spontaneous magnetic ordering transitions to ferromagnetic and antiferromagnetic states. The last chapter outlines the tensile, creep and fatigue properties of metals at low temperatures in terms of dislocation theory.

I am greatly indebted to my colleagues at Southampton for the many fruitful discussions I have had with them, and for their helpful advice and criticism of the text; particularly Dr. K. Kellner, Mr. S. Weintroub and Dr. E. M. Wray.

I am very grateful to the many authors who have kindly given permission to reproduce illustrations, and to Mr. A. J. Young for help with the other diagrams.

Acknowledgements

Grateful acknowledgement is made for permission to use illustrations which originally appeared in books and periodicals published by: American Institute of Physics, American Chemical Society, Institute of Physics and the Physical Society, Kamerlingh Onnes Laboratory, Oxford University Press, Royal Society, Taylor and Francis Ltd.

1 Introduction

1.1. Mass, Length, Time and Temperature

The world of the physicist is concerned largely with the understanding of the fundamental interactions in matter. His range of interest is closely linked with the magnitude of the physical dimensions of Mass, Length and Time, with which he works. In general, the fundamental interactions can be divided into two groups. Firstly, there are the elementary interactions of a small number of atomic or nuclear particles, either between themselves, or with electromagnetic radiation. Secondly, there are the collective interactions of a large number of atomic particles (atoms, electrons, molecules) in the form of a solid, a liquid, a non-ideal gas, or a plasma.

In this second group, a fourth physical dimension, that of *Temperature* is usually necessary to describe the collective interaction, and the physical behaviour of a collective system. In all fundamental interactions, whether elementary or collective, the allowed energy states are *quantised*, in discrete units or quanta of energy. Confining our attention to the behaviour of the collective solid state system of atoms and electrons, this energy quantisation can only be understood by ranging widely over the dimension of Temperature, away from the region of 300°K most familiar to us.

If we go up in temperature, then we are limited, since by the time we have covered one order of magnitude, or one ten-fold, increase in temperature to 3000°K, most solids have melted or vaporised. However, if we go down in temperature, we have many orders of magnitude available for studying the solid state.

Furthermore, in going down in temperature, the physical behaviour arising from energy quantisation, together with some unexpected forms of collective interaction, become revealed in many different and fascinating ways.

1.2. The Logarithmic Plot of Temperature

The linear scale of the absolute or Kelvin scale of temperature is very cramping in our thoughts about the relative importance of temperature intervals within the dimension of temperature. For example, the temperature range from the absolute zero of temperature 0°K, up to 1°K, appears insignificant compared with the range from 1°K up to 300°K.

A much more clear picture of the temperature spectrum is obtained by thinking in terms of the logarithm of the temperature, $\log_{10}$ T. The complete spectrum of temperatures attainable under laboratory conditions is shown in Figure 1.1, using such a logarithmic plot. The first point to note is that absolute zero becomes $10^{-\infty}$, in which form the principle of the unattainability of absolute zero may be better appreciated. The second point to note is that room temperature, 300°K or $10^{2.477}$°K, lies around the middle of the available temperature spectrum, between the region of low temperatures extending down to 10^{-6}°K, and the region of high temperatures extending up to 10^{7}°K.

The change in physical behaviour of a collective system has about equal significance, and is equally dramatic, over each order of magnitude change in temperature. We are all familiar with the changes that take place between 300°K and 3000°K; the deterioration in mechanical properties, the transformations to liquid and vapour phases, the loss of ferromagnetic properties, and so on. The changes in physical behaviour in going down in temperature by an order of magnitude, from 300°K to 30°K, are equally important; lattice specific heats decrease by

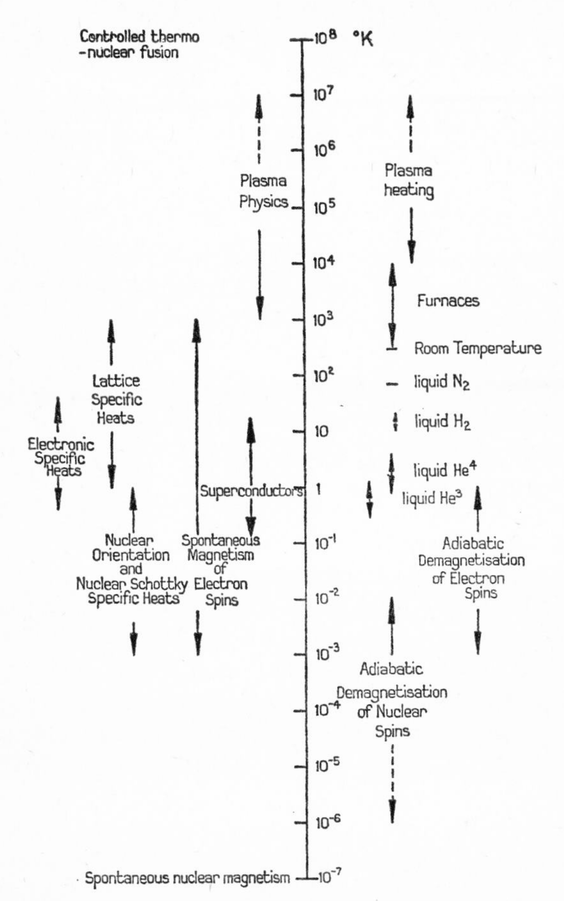

FIG. 1.1. Logarithmic plot of the temperature spectrum.

a factor of ten or more, electrical and thermal conductivities of pure materials increase by a factor of ten, and solids become stronger and/or more brittle. In going down the next two orders of magnitude in temperature, from 30°K to 0·3°K, we have, perhaps, a more interesting low temperature range, with the occurrence of the spontaneous order-disorder phenomena of superfluidity in liquid helium, and superconductivity in certain metals, alloys and compounds. Spontaneous ordering is a fundamental property of many, though not all, collective systems at low temperatures. The most well-understood ordering phenomena are the spontaneous magnetic transitions which all paramagnetic atoms (with permanent magnetic moments) undergo at sufficiently low temperatures. The whole temperature spectrum between 10^{-3}°K and 10^{3}°K is inundated with spontaneous magnetic transitions into the ferromagnetic, antiferromagnetic and ferrimagnetic states.

1.3. Absolute Zero, and the Third Law of Thermodynamics

The guiding principle to all physical behaviour at low temperatures is the Third Law of Thermodynamics. This states that the entropy, or disorder, of each system in internal thermal equilibrium, tends to zero at absolute zero.

Some important consequences follow from this simple statement. Firstly it is impossible to cool down to absolute zero. Secondly, many physical properties of a solid tend to vanish at absolute zero. These include thermal conductivities, thermal expansion coefficients, and all types of specific heat.

At absolute zero, each system attains a state of maximum order. However, there is no restriction by the Third Law to prevent a system from attaining its state of maximum order at a finite temperature. The spontaneous ordering phenomena of superconductivity, superfluidity and ferromagnetism are, in fact, natural short cuts to the perfectly ordered state at absolute zero.

We have already mentioned that a collective system possesses a discrete set of quantised energy levels. The state of maximum order corresponds to the occupation by the system of the lowest possible set of energy states. In this state, the energy of the system is a minimum, but is by no means zero. This minimum, or 'zero point', energy has an important bearing on many of the properties of the solid state at low temperatures.

1.4. Production of Low Temperatures

The starting point for achieving low temperatures down to about 1°K is the use of liquified gas as a refrigerant in a vacuum insulated dewar container. The main refrigerants include liquid helium He^4 with a normal boiling point, under a pressure of 760 mm Hg, at 4·2°K, liquid hydrogen at 20·4°K, liquid nitrogen at 77°K, liquid oxygen at 90°K and carbon dioxide freezing mixture at 195°K.

The production of temperatures away from the normal boiling point of a refrigerant may be achieved in two ways:—

1. By variation of the vapour pressure above the liquid. Reduction of the vapour pressure, by pumping, results in a lowering of the temperature of the liquid below the normal boiling point. Except for He^3 and He^4 which remain liquid under their saturation vapour pressure down to 0°K (Figure 1.2(a)), the lower limit is set by the triple point, below which only the solid is in equilibrium with the vapour.
   In comparison with liquids, solidified gases are unsatisfactory refrigerants on account of their low thermal conductivity and poor internal thermal equilibrium. The lowest temperatures attainable by pumping liquid He^3 or He^4 are limited by the speed of the pumping systems. The temperature ranges generally available by vapour pressure reduction are 0·3°K–3·2°K for He^3 (only small and expensive quantities available), 0·9°K–4·2°K for He^4, 14°K–20·4°K for hydrogen, and 63°K–77°K for nitrogen. The ranges can

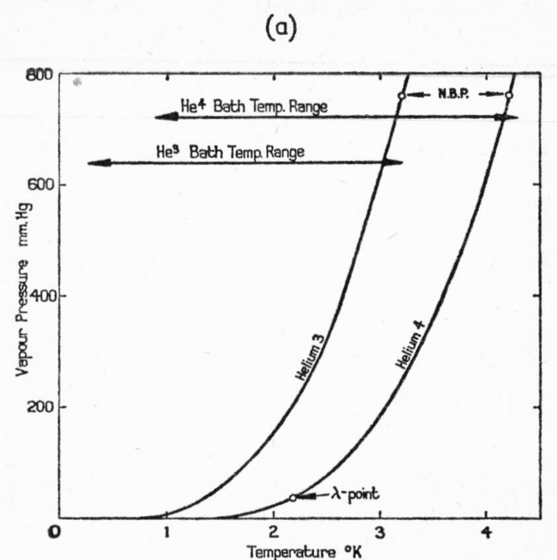

FIG. 1.2(a). Vapour pressure-temperature curves of He^3 and He^4 showing temperature ranges available by pumping.

be extended above the normal boiling points by pressurizing above 760 mm Hg, but this is usually inconvenient.

2. By using the principle of the adiabatic calorimeter to attain temperatures above the normal boiling point of the refrigerant (Figure 1.2(b)). The metal working chamber, with heater coil attached to it, is insulated from the refrigerant by a second evacuated chamber surrounding it. By passing an electric current through the heater coil, the temperature of the working chamber may be raised and conveniently controlled above the temperature of the refrigerant.

For attaining very low temperatures below 1°K, down to about 0·001°K, it is necessary to use a completely different method of refrigeration, namely the adiabatic demagnetisation of a paramagnetic salt. The principle of this magnetic cooling

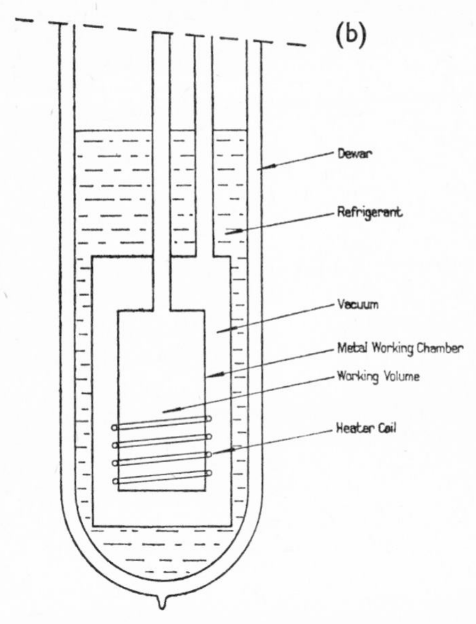

FIG. 1.2(b). Schematic drawing of adiabatic calorimeter for attaining temperatures above normal boiling point of refrigerant.

method may be illustrated with the entropy-temperature diagram of a suitable salt (Figure 1.3(a)).

At the normal starting temperature of 1°K, achieved by pumping the liquid He^4 bath (Figure 1.3(b)), the entropy S_L of the lattice is small in comparison with the magnetic entropy of the paramagnetic spin system. At lower temperatures, the magnetic entropy S_0 in zero magnetic field falls off steeply, when spontaneous magnetism of the atomic spins takes place, setting a limit to the lowest temperature attainable.

When a large magnetic field, 10–20 kOe is applied to the salt at 1°K, the lower magnetic levels become preferentially thermally populated, and the entropy S_H, of the magnetic spins in the field, falls from A to B. Once the heat of magnetisation, assoc-

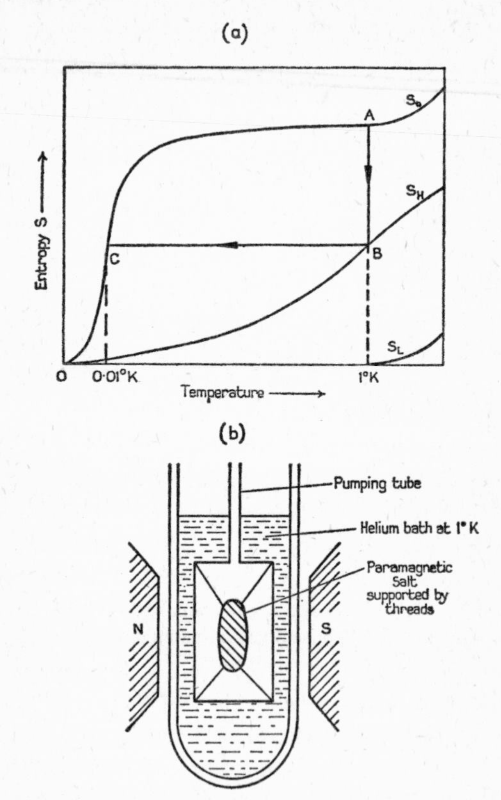

FIG. 1.3(a). Entropy-temperature diagram of a paramagnetic salt, illustrating cooling by adiabatic demagnetisation.

FIG. 1.3(b). Schematic diagram of apparatus for cooling by adiabatic demagnetisation.

iated with this decrease in entropy, $\Delta Q = T(S_0 - S_H)$, has been conducted away to the helium bath, the salt is thermally isolated by pumping away the helium exchange gas surrounding it. On reducing the magnetic field to zero adiabatically, the entropy remains constant, and the temperature of the salt falls from

B to *C*, reaching about 10^{-2}°K. The salt then warms up, back to the helium bath temperature at 1°K, at a rate depending on the heat influx, and the specific heat of the salt.

1.5. Measurement of Low Temperatures

The practical measurement of temperature usually depends on the use of one or other form of resistance thermometer, chosen to have a suitably large and reproducible temperature coefficient of resistance in the temperature range of interest. There is no single satisfactory thermometer which covers the whole range between 1°K and 300°K. Platinum resistance thermometers are applicable down to 20°K or possibly 10°K, suitably doped semiconductor resistance thermometers may be used between about 2°K and 100°K, and certain carbon radio resistors are suitable between 0·1°K and about 20°K.

The thermo-electric power of most metals becomes too small for thermo-couples to be used for *accurate* measurement of low temperatures. However, thermo-couples find wide use as simple and convenient temperature indicators. The copper-constantan combination is useful down to 20°K, while gold alloys containing small amounts of cobalt, when used with copper, are applicable down to liquid helium temperatures.

Below 1°K, temperatures are usually determined from the magnetic susceptibility of paramagnetic salts (see section 5.5).

Calibration of practical thermometers against primary constant volume gas thermometers is very laborious. The usual practice is to calibrate against easily reproducible, secondary thermometer scales, like the vapour pressure-temperature scales of pure He^3, He^4, the equilibrium composition of ortho- and para-hydrogen, and nitrogen, and the platinum resistance thermometer scale above 90°K. At the lower temperatures, extrapolation of the calibration is necessary over the gaps between the ranges covered by vapour-pressure thermometers.

2 The Specific Heat

2.1. Introduction

The specific heat C_p of one mole (gram molecular weight), of a solid at constant pressure, is determined experimentally from the relation

$$C_p = \frac{\Delta Q}{\Delta T} \text{ cal deg}^{-1} \text{ mole}^{-1}$$

where ΔQ is the quantity of heat required to produce a small change in temperature ΔT. Thus, specific heat measurements are simple in principle, and suitable calorimetric techniques are available for covering temperature ranges from 0·01°K up to 300°K, and higher. The measurements are repetitive by their nature, but the labour can be considerably reduced by intelligent use of computer programming.

Specific heat determinations are important for two broad reasons. Firstly, the specific heat is closely related to the thermodynamic coordinates of a solid. The third law of thermodynamics predicts that, for all types of phenomena which maintain internal thermal equilibrium, the associated specific heats become zero at the absolute zero of temperature. Thus, specific heat measurements over temperature ranges, commencing from temperatures close to 0°K, provide a record of the thermal history of the solid. For example, the internal thermal energy at constant volume is given by

$$U_v = \int_0^T C_v \, dT \tag{2.1}$$

and the entropy by $$S = \int_0^T \frac{C_v}{T} dT \tag{2.2}$$

Secondly, the temperature variation of the specific heat provides a simple basis for testing physical models of the associated phenomena. It should be noted that specific heats do not, in themselves, tell us anything about the physical nature of the phenomena.

It is important to realize the difference between the two specific heats, C_p and C_v, applicable to constant pressure and constant volume conditions, respectively. The experimental quantity is normally C_p, whereas the theoretically derived quantity is C_v. The two are related thermodynamically by

$$C_p - C_v = \beta^2 \eta V T$$

where β is the volume expansion coefficient, η is the bulk modulus or reciprocal of the compressibility, and V is the molar volume of the sample. The difference between C_p and C_v is negligibly small at liquid helium temperatures, but becomes progressively larger, and must be taken into account, at higher temperatures.

2.2. The Collective Motion of an Atomic Lattice

At room temperature, the molar heat, or specific heat of one mole, of a solid element is approximately constant, equal to $3R$. The solid can be satisfactorily described by classical physics, with each atom pictured as vibrating in 3 dimensions about a fixed point, with an energy of $3kT$ according to the 'equipartition of energy' theorem. The total energy of one mole containing N atoms is $3NkT$ or $3RT$ and the molar heat has the constant value of $3R$ = 5·96 cal mole^{-1} deg^{-1} = 24·94 joules mole^{-1} deg^{-1} (N is Avogadro's number, R is the gas constant).

However, at low temperatures, the lattice specific heat falls in a regular manner with decreasing temperature, approaching zero at a rate proportional to T^3 below about 20°K. (Figure 2.1). This behaviour cannot be explained by classical physics. It arises firstly from the *collective vibrations* of the atoms in the solid crystalline lattice, and secondly from the *quantum nature* of these collective vibrations.

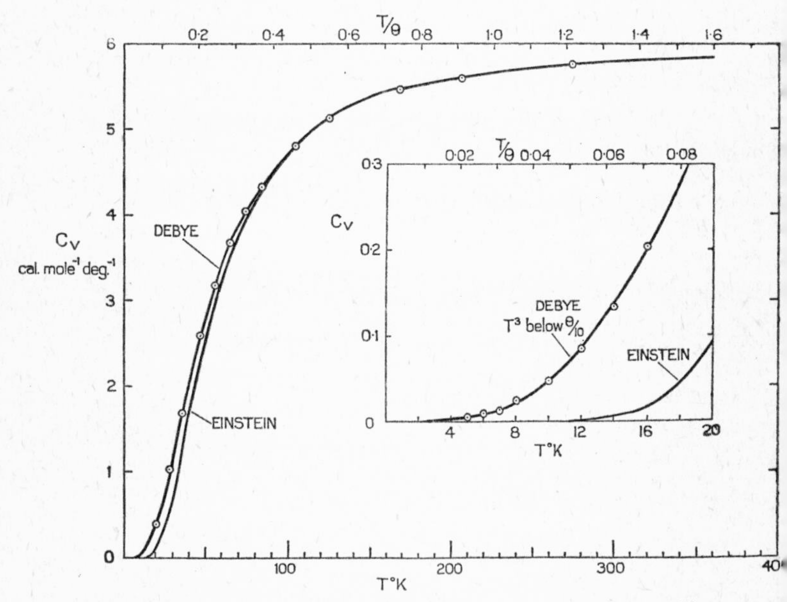

FIG. 2.1. Comparison of experimental values of the specific heat for silver with values calculated on the Einstein and Debye models.

The thermal vibrations of atoms in a lattice are *not* random and independent of one another because the atoms are closely linked together by elastic interatomic forces. These forces are responsible for the regular arrangement of atoms in a crystal lattice, and for the precise geometrical shape of single crystals. The close linkage allows collective motion of the atoms to be thermally excited in the form of *elastic waves*. These elastic waves in the solid medium behave in much the same way as sound waves, apart from their very much higher frequencies (up to 10^{13} c/s). The complex thermal motion of each atom is, in fact, made up from the superposition of a large number of these waves travelling in all directions through the 3-dimensional lattice with approximately the velocity of sound.

The complex motion can be best pictured by singling out for consideration the thermally excited wave motion of a 1-dimensional chain of N atoms, and summarizing the relevant properties:—

1. At thermal equilibrium, only standing waves need to be considered. Each standing wave is composed of two travelling waves moving in opposite directions with approximately the velocity of sound in the lattice.

2. In a chain of N atoms, a total of $3N$ independent standing waves or normal modes may be excited. There are N modes for longitudinal waves, producing motion of the atoms parallel to the direction of propagation along the chain, and $2N$ plane polarised modes for transverse waves, which produce motion perpendicular to the chain.

3. To a good approximation, the standing waves produce simple harmonic motion of the atoms, and are therefore independent of one another. Slight anharmonic motion does in fact take place. This allows some interaction to take place between the standing waves so that (a) thermal equilibrium can be achieved, and (b) a scattering mechanism can exist to limit heat conduction by the lattice waves (see section 3.3).

4. At the same time as there is a natural limit to the number ($3N$) of distinct standing waves, there is also a natural upper limit to the frequency of the waves, called the cut-off

frequency ν_{max}. The chain (or lattice) behaves as a low-pass filter, and waves with frequencies greater than ν_{max} cannot pass. The corresponding lower limit to the wavelength, λ_{min}, is just twice the interatomic spacing, when adjacent atoms are moving 180° out of phase with each other.

In an atomic lattice with an interatomic spacing of 10^{-8} cm, λ_{min} is 2×10^{-8} cm. The sound velocity in a solid is about 3×10^5 cm sec^{-1}, so that the natural cut-off frequency ν_{max} is approximately 10^{13} c/s. This frequency is very much higher than the maximum ultrasonic frequency of 10^9 c/s which can be generated at present in the laboratory using quartz. The corresponding frequency of electromagnetic radiation lies in the infra-red region.

5. The energy associated with each standing wave is quantised, in the sense that the energy is only allowed to consist of discrete values E_n which are related to the frequency ν by

$$E_n = (n + \tfrac{1}{2})h\nu \quad \text{where } n = 0, 1, 2, 3 \text{ etc.} \tag{2.3}$$

and h is Planck's constant.
This energy quantisation is a fundamental difference in behaviour between collective systems on the atomic and macroscopic scales which will always be met in the physics of solids at low temperatures.

6. According to the Boltzmann distribution law, the population of any quantised energy level E_n is proportional to $\exp -E_n/kT$. Thus the *average* energy of each mode is given by

$$E(\nu) = \frac{\sum_{n=0}^{\infty} (n + \frac{1}{2})h\nu \exp[-(n + \frac{1}{2})h\nu/kT]}{\sum_{n=0}^{\infty} \exp[-(n + \frac{1}{2})h\nu/kT]}$$

which reduces to

$$E(\nu) = \tfrac{1}{2}h\nu + \frac{h\nu}{\exp[h\nu/kT] - 1} \tag{2.4}$$

The first term is the zero-point energy E_0, possessed by the mode in its lowest ($n = 0$) energy state. At the absolute zero of temperature, when only the lowest energy state is occupied, the atoms will still have a finite motion, and a finite zero-point energy. The second term is the thermal energy, and is, of course, temperature dependent.

For a 3-dimensional lattice of N atoms, the idea of normal modes carries straight over from the 1-dimensional chain. There is a total of $3N$ modes and a cut-off frequency ν_{max} arises quite naturally again as a property of the lattice.

The total energy U of the lattice is the sum of the average energies $E(\nu)$ of all the $3N$ modes within the frequency range between 0 and ν_{max}.

i.e.
$$U = \int_0^{\nu_{max}} f(\nu)E(\nu)\, d\nu \tag{2.5}$$

where $f(\nu)$ is the frequency distribution of the modes, or the number of modes per unit range of frequency.

The specific heat C_v is then obtained from the relation

$$C_v = \frac{dU}{dT}$$

The major difficulty lies in determining the frequency distribution $f(\nu)$ of the $3N$ modes over the whole frequency range up to ν_{max}. Two simple distributions, which meet with considerable success in accounting for the temperature variation of the lattice specific heat of solids, are those of the Einstein and Debye models. (Figures 2.2(a) and (b)).

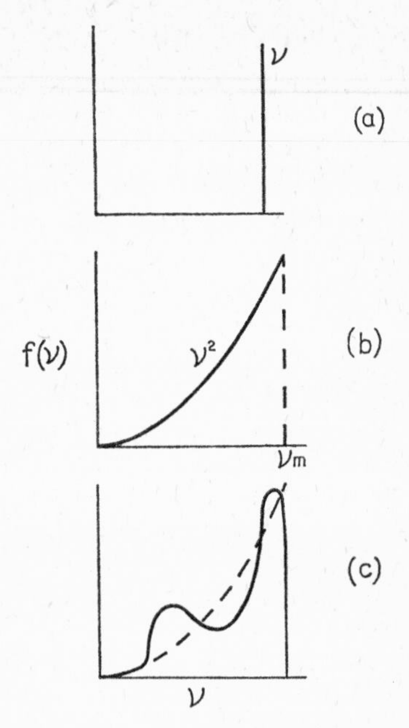

FIG. 2.2. The frequency spectrum of lattice vibrations in a solid according to (a) Einstein, (b) Debye, and (c) Blackman.

2.3. The Einstein Model of Lattice Specific Heats

Einstein proposed a model in which the vibrations of a 3-dimensional lattice of N atoms are treated as a set of $3N$ independent, one dimensional, simple harmonic oscillators, each having the *same frequency*. Einstein's oscillators are completely equivalent to $3N$ of the normal modes described above, all with the same frequency ν. The value of ν is chosen to give the best fit with experiment, and turns out to be quite close to ν_{max}.

On the Einstein model, the total energy U_E from equation (2.5) reduces to

$$U_E = 3NE(\nu)$$

$$= \frac{3Nh\nu}{2} + \frac{3Nh\nu}{\exp(h\nu/kT) - 1}$$

The first term is temperature independent and is the zero-point energy. The Einstein specific heat at constant volume is given by

$$C_E = \frac{dU_E}{dT}$$

$$= 3Nk\left(\frac{h\nu}{kT}\right)^2 \frac{\exp(h\nu/kT)}{[\exp(h\nu/kT) - 1]^2} \qquad (2.6)$$

At high temperatures ($kT \gg h\nu$),

$$C_E = 3Nk = 3R, \text{ the classical value.}$$

At low temperatures ($kT \ll h\nu$)

$$C_E = 3R\left(\frac{h\nu}{kT}\right)^2 \exp(-h\nu/kT)$$

$$\approx 3R \exp(-h\nu/kT) \text{ in t e limit}$$

at the lowest temperatures.

This exponential variation of C_E is much more rapid than the observed T^3 relation, at the lowest temperatures. (Figure 2.1). At the higher temperatures, however, a careful choice of ν leads to quite good agreement between C_E and experimental values.

2.4. The Debye Model of Lattice Specific Heats

It is clear from our previous discussion of a 1-dimensional chain, that the normal modes of vibration extend over a finite frequency range, and that the choice of one frequency on the Einstein model is an over-simplification. On the Debye model, the frequency spectrum of a monatomic solid is obtained by treating the solid as a homogeneous, continuous medium. The velocities, v_T and v_L, of transverse and longitudinal sound waves are assumed to apply right up to a maximum frequency, and to be independent of propagation direction (i.e. the medium is elastically isotropic). Taking into account both transverse and longitudinal normal modes, the frequency distribution $f(\nu)$ is a parabolic function given by

$$f(\nu) = 4\pi\left(\frac{2}{{v_T}^3} + \frac{1}{{v_L}^3}\right)\nu^2 \tag{2.7}$$

The maximum frequency ν_m is determined by integrating $f(\nu)$ over the $3N$ modes with the lowest possible frequencies.

i.e.
$$\int_0^{\nu_m} f(\nu)\, d\nu = 3N$$

This definition of ν_m is quite close to, but *not the same* as, the natural cut-off frequency $\nu_{\max}$ of the lattice. Carrying out this integration, ν_m is given by

$${\nu_m}^3 = \frac{9N}{4\pi}\Bigg/\left(\frac{2}{{v_T}^3} + \frac{1}{{v_L}^3}\right) \tag{2.8}$$

Substituting equations (2.4), (2.7) and (2.8) into equation (2.5,) and ignoring the temperature independent, zero point energy term, the temperature dependent energy is

$$U = \frac{9N}{{\nu_m}^3}\int_0^{\nu_m} \frac{h\nu^3}{\exp(h\nu/kT) - 1}\, d\nu$$

Writing $k\theta = h\nu_m$, we introduce a characteristic or Debye temperature θ, and obtain

$$U = \frac{9NkT^4}{\theta^3}\int_0^{\theta/T} \frac{x^3}{(e^x - 1)}\, dx$$

where
$$x = \frac{h\nu}{kT}$$

The Debye specific heat at constant volume is then

$$C_v = 9R\left(\frac{T}{\theta}\right)^3 \int_0^{\theta/T} \frac{x^4 e^x}{(e^x - 1)^2}\, dx \tag{2.9}$$

This specific heat function is plotted in Figure 2.1 together with experimental data on silver.

At high temperatures ($T > \theta$), the Debye specific heat function approaches the classical value of $3R$ mole^{-1}. At low temperatures ($T \ll \theta$),

$$C_v = 234R\left(\frac{T}{\theta}\right)^3 = 464{\cdot}4\left(\frac{T}{\theta}\right)^3 \text{ cal mole}^{-1} \text{ deg}^{-1} \tag{2.10}$$

The difference between equations (2.9) and (2.10) at a temperature of $\theta/10$ is about 2%. At lower temperatures, the difference becomes rapidly smaller, and equation (2.10) becomes a simple, convenient and accurate description of the lattice specific heat.

2.5. Limitations of the Debye Model

In general, there is quite reasonable agreement, for many solids, between the Debye specific heat curve and experimental measurements of the lattice specific heat C_v. Values of θ, for some of the elements, are given in Table 2.1. However, close examination reveals somewhat similar deviations in all cases.

Debye's assumption of an isotropic medium is very good at low temperatures ($T < \theta/10$), where only the low frequency modes are thermally excited. These modes have wavelengths many times the atomic spacing and so do not 'see' the atomicity of the lattice. Hence, the T^3 variation, of C_v in equation 2.10, fits experimental data very closely. However, at higher temperatures, the shorter wavelength modes, which become thermally excited, do 'see' the atomicity of the lattice, and the Debye frequency distribution is modified.

TABLE 2.1.

	θ°K		θ°K
Ag	225	Li	369
Al	426	Mg	342
Au	164	Mn	450
Be	1160	N	68
Bi	100	Na	158
C (diamond)	2200	Ni	440
Cl	115	Pb	108
Co	443	Pt	221
Cs	40	Sn (grey)	212
Fe	464	Ti	430
Ge	363	U	200
Hg	75	Zn	310

Values of the Debye characteristic temperature, θ, for some elements

Blackman and others have made a detailed analysis of the modes of vibration to be expected in 3-dimensional crystals, and have deduced frequency distributions differing considerably from the simple parabolic form of Debye's theory (Figure 2.2(c)). These distributions coincide with the Debye distribution at low frequency, but at higher frequencies, they rise more steeply to a complicated double-humped distribution, differing widely for different crystal structures.

Although the Debye specific heat function does not give an accurate description at temperatures above $\theta/10$, it is a good guide to the general behaviour of lattice specific heats. Consequently, it has become customary to indicate the non-ideal behaviour of the lattice specific heats in the following way. The measured lattice specific heat at each temperature T is marked on the standard Debye specific heat curve (equation (2.9) or Figure 2.1) and gives a corresponding value of θ/T. Since T is known, an 'effective' θ is determined.

It must be remembered that θ has then lost its significance as a measure of the cut-off frequency. The general trend, of the variation of the 'effective' θ with T (Figure 2.3) shows a temperature independent region θ_0 at the lowest temperatures, where

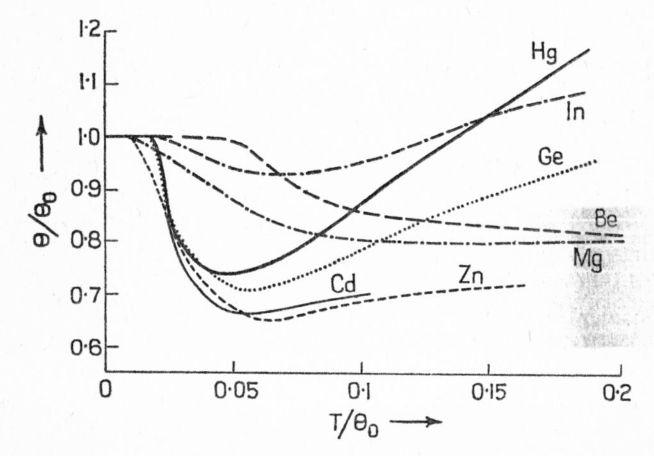

FIG. 2.3. Variation of θ with temperature for different solids. θ_0 is the value of θ at $T = 0$°K. (after Parkinson, 1958)

the T^3 variation in C_v is accurately described. At higher temperatures, there is a sharp dip, corresponding to a rise of the frequency distribution curve above the parabolic Debye distribution.

2.6. Phonons

The description of lattice vibrations by a distribution of normal modes, each possessing a set of quantised energy states, is quite satisfactory for describing lattice specific heats. To understand the other thermal properties of the lattice, we need a picture of the 'quanta' of energy $h\nu$, which the lattice waves can possess.

Every normal mode or standing wave is composed of two travelling waves, moving in opposite directions with approximately the velocity of sound. Each quantum of energy $h\nu$, taken up by a standing wave as a result of thermal excitation, can be regarded as a wave packet, or wave train of finite extent, associated with one, or other, of the two constituent travelling waves. This wave packet is then propagated along the standing wave, or through the lattice, with the group velocity of the travelling wave, and is called a *phonon*. Energy quantisation corresponds to each normal mode possessing an integral number, only, of phonons with energy $h\nu$, at every instant of time.

Phonons carry energy and momentum, and are scattered by other phonons, conduction electrons, lattice boundaries and lattice imperfections. These particle-like properties enable us to consider them as particles in several applications. For example, we shall use the particle concept of phonons when we consider lattice heat conduction in Chapter 3.

2.7. Conduction Electrons in Metals

When atoms are assembled in a crystal lattice to make a metal, the valence electrons become detached from their parent atoms and move through the lattice like an electron gas, confined only by the bounding surfaces of the metal. If this electron gas is considered as a classical system obeying Maxwell-Boltzmann statistics, like the molecules of a gas in a container, then we should expect each electron to have an average kinetic energy of $(\frac{3}{2})kT$. Assuming one valence electron per atom, the associated electronic specific heat would be $(\frac{3}{2})R$, independent of temperature, and hence finite at 0°K, in contradiction to the Third Law of Thermodynamics. Furthermore, we should expect, at high temperatures ($T > \theta$), a total specific heat of $3R$ for insulators, and a value of $(3 + \frac{3}{2})R$ or $(\frac{9}{2})R$ for metals.

In fact, the specific heat of metals, at high temperatures, is only slightly greater than that for insulators. In other words, the electronic specific heat is small compared with lattice specific heats, and cannot be explained on a classical picture using Maxwell-Boltzmann statistics.

This difficulty was not overcome, until it was realised that the electron gas obeys a different type of statistics, namely quantum or Fermi-Dirac statistics. On the classical theory, the kinetic energy of a particle can have any value out of a continuous distribution of energy values, although, of course, some values are more probable than others. On the quantum mechanical picture, the continuous distribution of available energies is replaced by a discrete set of allowed quantum energy levels (each of which can be held by a maximum of two particles, in the case of electrons). For real gases, the spacing between adjacent levels is extremely small, and the number of particles is very much smaller than the available quantum states. Consequently, the difference between the quantum mechanical and the continuous classical distribution is so small, that no deviation from the classical behaviour of real gases is observed, even at the lowest temperatures.

However, for electrons in metals, the situation is completely different, and quantum mechanical behaviour is shown at all temperatures. The spacing between levels is larger, because of the small electron mass, and the density of free electrons is so high, that the number of electrons is comparable with the number of available energy states. The latter is emphasised by the role of the Pauli exclusion principle, which states that no two electrons, in a given system, can have the same set of quantum numbers, at the same time. Since the electron has an intrinsic spin angular momentum of $\frac{1}{2}(h/2\pi)$, and two spin directions, the effect of the Pauli exclusion principle is that only two electrons can occupy the same translational energy level. It follows immediately that, at absolute zero, the kinetic energy of the free electrons cannot be zero, since this would mean that all the electrons had one particular energy. In fact, the electrons occupy the lowest possible set of energy levels, consistent with the Pauli principle, and their mean energy, or zero-point energy, is very far from zero.

At finite temperatures, the additional thermal energy of the electrons is very small in comparison with the zero-point energy, and the electronic specific heat is small. The distribution of the electrons, among the available energy states, can only be found by using Fermi-Dirac statistics, which take the Pauli exclusion principle into account.

2.8. The Fermi Gas Model of Conduction Electrons in Metals

The Fermi gas model is by no means a complete description, but it serves to illustrate the quantum mechanical behaviour of conduction electrons in metals. The electrons are assumed to be completely unbound, and confined only by the metal boundary. Taking their wave-like properties into account, wave mechanics shows that the electrons can occupy a set of discrete energy levels. (cf. the crystal lattice, with a discrete number of vibrational modes, each with a discrete set of energy levels). The number of levels per unit volume of metal, $N(W)$, with energies between W and $W + dW$ is given by

$$N(W)\,dW = 4\pi\left(\frac{2m}{h^2}\right)^{\frac{3}{2}} W^{\frac{1}{2}}\,dW \tag{2.11}$$

At absolute zero, all the lowest possible states will be occupied up to a maximum energy, W_F, called the 'Fermi energy'. Then the number of conduction electrons per unit volume, n, is equal to the integral of the distribution function $N(W)$, over the energy range between zero and W_F, i.e.

$$n = \frac{8\pi}{3}\left(\frac{2m}{h^2}\right)^{\frac{3}{2}} W_F^{\frac{3}{2}} \tag{2.12}$$

Thus for a Fermi gas, W_F depends only on n.

Combining equations (2.11) and (2.12,) the distribution function becomes

$$N(W)\,dW = \frac{3nW^{\frac{1}{2}}}{2W_F^{\frac{3}{2}}}\,dW \tag{2.13}$$

Plotting $N(W)$ against W, a parabola is obtained, with a sharp cut off at the Fermi energy W_F, when $T = 0°K$ (Figure 2.4). This upper limit to $N(W)$, at W_F, is generally called the *Fermi surface* of the electron distribution.

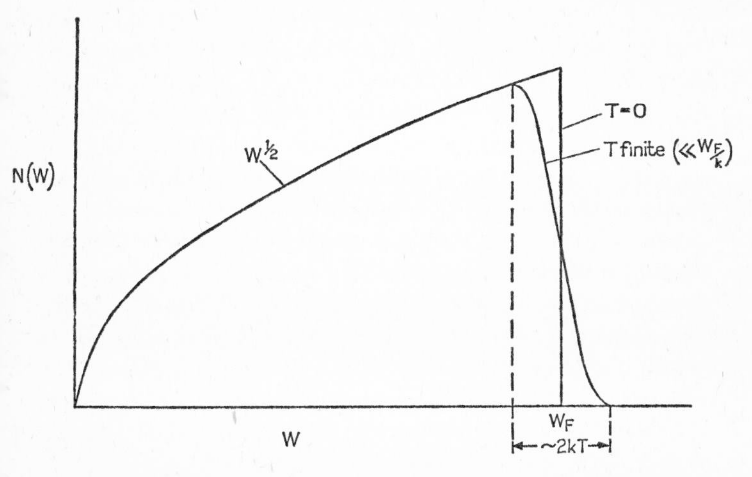

FIG. 2.4. The electron density of states $N(W)$ as a function of energy W, for $T = 0°K$ and $T \ll W_F/k$. Only a relatively small region, within kT of the Fermi energy W_F, is modified when the temperature is increased.

The mean energy $\overline{W}$ of the electrons at absolute zero, can be found by integrating over the allowed energy states,

i.e.
$$\overline{W} = \frac{1}{n}\int_0^{W_F} W \,.\, N(W)\,dW \tag{2.14}$$

Substituting for $N(W)$ from equation (2.13,)

$$\overline{W} = \frac{3}{2} W_F^{-\frac{3}{2}} \int_0^{W_F} W^{\frac{3}{2}}\,dW$$
$$= \frac{3}{5} W_F$$

Thus the total zero point energy U_0 is equal to $\frac{3}{5}nW_F$ per unit volume.

Taking sodium as a typical example, and assuming one free electron per atom, then $n = 2{\cdot}5\ 10^{22}\ \text{cm}^{-3}$ and $W_F = 3{\cdot}1$ eV. The mean energy $\overline{W}$ is 1·9 eV and the zero point energy U_0 is 2,500 cal cm^{-3} or 60,000 cal mole^{-1}. This is a very large quantity, and equivalent to a classical thermal energy of $\frac{3}{2}RT$ at a temperature of 20,000°K.

At finite temperatures, the electrons tend to acquire thermal energy of the order of kT, in addition to their large zero point energy. However, by the Pauli principle, only those electrons within the energy range, $(W_F - kT)$ to W_F, can be thermally excited into empty allowed states above W_F. Thermal excitation of electrons with lower energies is prevented, since there are no empty higher energy levels, within a range of kT, for them to occupy. The thermal modification of the distribution, among the states with energies close to W_F, can be described using the proper Fermi-Dirac statistics.

Quoting the result for a Fermi gas in thermal equilibrium at temperature T, the distribution per unit volume is given by

$$N(W)\,dW = \frac{CW^{\frac{1}{2}}dW}{e^{(W-W_F)/kT}+1} \tag{2.15}$$

where
$$C = \frac{3n}{2W_F^{\frac{3}{2}}}$$

The distributions for $T = 0°K$, and $T \ll W_F/k$ (or $T \ll 20{,}000°K$) are shown in Figure 2.4. As T increases, the step in the distribution at W_F rounds off; levels within about kT *below* W_F are partially depopulated, while empty levels within kT *above* W_F become partially populated.

Substituting equation (2.15) in (2.14), the total energy U per unit volume is given by

$$U = \int_0^\infty \frac{CW^{\frac{3}{2}}\,dW}{e^{(W-W_F)/kT}+1}$$

$$= U_0 + \frac{\pi^2}{6}(kT)^2 N(W)_F$$

+ higher order terms in T (negligible).

Hence the electronic specific heat per unit volume is

$$C_e = \frac{dU}{dT} = \frac{\pi^2}{3}k^2T\,.\,N(W)_F \qquad (2.16)$$

and varies linearly with temperature.

$N(W)_F$ is the density (i.e. number per unit interval of energy) of allowed states at the Fermi surface energy W_F. More conveniently, C_e is written as

$$C_e = \gamma T, \text{ where } \gamma = \frac{\pi^2}{3}k^2N(W)_F$$

Substituting the value of $N(W)_F$ for a Fermi gas from equation (2.13)

$$C_e = \frac{n\pi^2k^2T}{2W_F} \text{ per unit volume.} \qquad (2.17)$$

On comparison with the classical value, $C_e = \frac{3}{2}nk$, it can be seen that the quantum statistical value of the electronic specific heat is smaller, by a factor of the order of $\frac{kT}{W_F}$; about 100 times smaller at $T = 300°K$. This arises because only a fraction, of the order of $\frac{kT}{W_F}$, of the conduction electrons are able to increase their energy by thermal excitation.

Thus the difficulty, presented by the prediction of a large electronic specific heat by classical theory, is removed.

2.9. Experimental Measurements on Electronic Specific Heats

The electronic specific heat of metals is so small, that it can only be determined at low temperatures. For temperatures less than $\theta/10$ (about 20°K, generally), the lattice specific heat is proportional to T^3, and falls off very rapidly with decreasing temperature, eventually becoming smaller than the electronic contribution, which varies with T. At these temperatures, the total specific heat of a metal can be written in the form

$$C_v = aT^3 + \gamma T \tag{2.18}$$

Experimentally, it is convenient to rewrite equation (2.18) as

$$C_v/T = aT^2 + \gamma$$

and plot C_v/T against T^2. A straight line graph is then obtained, where the slope is 'a', and the intercept gives the value of γ. The experimental values obtained for silver, are shown plotted in this way in Figure 2.5. Deviations from the straight line plot will be observed at higher temperatures, as the lattice specific heat deviates from the T^3 law.

Experimental values of the electronic specific heat coefficient γ, for a number of metals, are compared in Table 2.2 with the values calcuated from the free-electron Fermi gas model. Although the observed and calculated values are of the same order of magnitude, there is, in general, no close agreement.

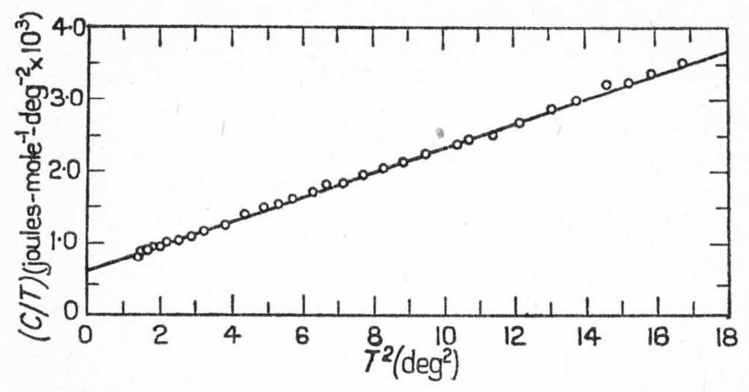

FIG. 2.5. The low temperature specific heat of silver, plotted as C/T against T^2. The intercept, when $T^2 = 0$, gives γ, the electronic specific heat coefficient. (from Corak, Garfunkel, Satterthwaite and Wexler, 1955)

TABLE 2.2

Element	Valency	W_F (eV) (from eqn 2.12)	$\gamma \times 10^4$ joules mole^{-1} deg^{-2} (experimental)	(theoretical)
Na	1	3·12	13·7	22·8
Cu	1	7·04	7·44	4·95
Ag	1	5·51	6·09	6·0
Be	2	13·8	2·22	4·8
Al	3	11·5	13·6	8·5
Pb	4	9·4	33·6	16.0
Co	1·6	9·2	47·5	3·4
Fe	2·1	11·5	50·2	4·2
Mn			180	
Pt	0·6	4·2	66·3	5·10

Values of γ, the electronic specific heat coefficient, obtained from experiment and calculated from equation (2.17)

In particular, the observed values of γ for transition metals are considerably higher. To understand this discrepancy, we must re-examine the model of a free-electron gas in metals.

2.10. Occurrence of Energy Bands

In the Fermi gas model, the conduction electrons are assumed to be bound to the metal as a whole, and not to individual atoms. They, therefore, move in a rectangular potential well as shown in Figure 2.6(a). On the other hand, for insulators,

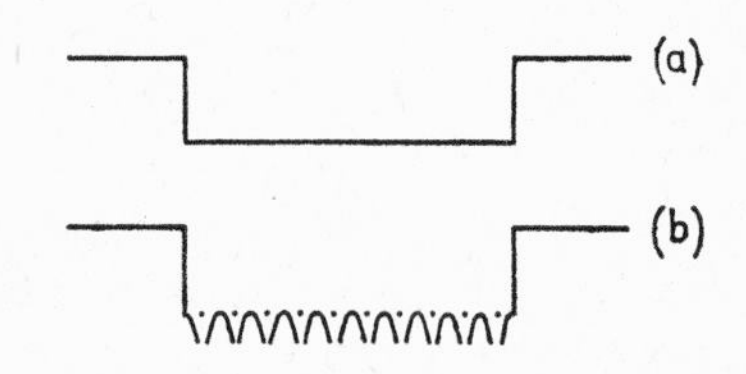

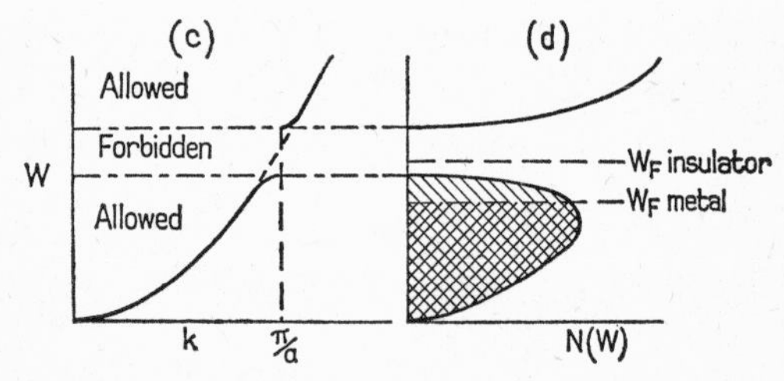

FIG. 2.6. (a) Potential energy assumed in free electron (Fermi gas) model of a metal

(b) Actual variation in potential energy, showing sharp fall near each atom.

(c) Plot of W against k, showing band structure due to periodic variation in potential of the lattice.

(d) Associated band structure of the density of states $N(W)$.

it is assumed that there are no free electrons, all electrons being firmly bound to their parent atoms. The intermediate class of solids, known as semi-conductors, which exhibit small electrical conductivities and generally a negative temperature coefficient of resistivity, cannot be explained at all on the free electron model.

To understand these different classes of solids, we must consider the behaviour of atomic electrons more closely, when the atoms are packed in a solid structure. We find that the potential energy of a conduction electron is not constant, but varies rather as shown in Figure 2.6(b), falling steeply when the electron is near each positively charged atom, and therefore varying with the periodicity of the lattice. The motion of electrons in such a potential is very complicated, and will only be discussed qualitatively here.

It is now well established from experiments with electron, neutron and other particle beams, that a moving particle has wave-like properties. The wavelength λ, associated with a particle of linear momentum **p** is given by the de Broglie relation,

$$\lambda = h/p = \frac{2\pi}{k}$$

Alternatively, $\mathbf{k} = \mathbf{p}\frac{2\pi}{h}$, where **k** is the 'wave vector' parallel to the direction in which the particle is travelling. For a free electron, with kinetic energy E electron volts, the de Broglie wavelength is given numerically by

$$\lambda = \sqrt{\frac{150}{E}} \times 10^{-8}\text{cm.}$$

Some representative values are given in Table 2.3.

TABLE 2.3.

Energy (eV)	$\lambda \times 10^{-8}$ cm
1	12·3
4	6·15
9	4·1

De Broglie wavelength of low energy, free electrons

Now X-rays are strongly diffracted in a crystal, when the wavelength satisfies the condition for constructive interference of reflected rays from adjacent Bragg planes of atoms,

$$n\lambda = 2d \sin \theta$$

where n is an integer, 'd' is the spacing of the Bragg planes and θ is the angle between Bragg plane and incident wave. Since 'd' is approximately equal to the inter-atomic spacing 'a', it is seen that strong diffraction, or Bragg reflection, can be expected only when λ is of the order of $2a$. Longer wavelength X-rays will be transmitted without Bragg reflection.

The condition for Bragg reflection of conduction electrons is closely similar. In a metal crystal with an atomic spacing of, say, 2×10^{-8} cm, low energy electrons (~ 1 eV, $\lambda \sim 12 \times 10^{-8}$ cm) will be transmitted. Bragg reflections start to take place when their energy approaches 9 eV, or as the de Broglie wavelength decreases to 4×10^{-8} cm, or as the electron wave vector **k** approaches the value π/a (when $\lambda = 2a$, $k = 2\pi/\lambda = \pi/a$). Now Bragg reflections give rise to stationary standing wave states, which do not fulfil the travelling wave requirement of conduction electrons. Consequently, bands of energy arise, which are forbidden to conduction electrons, and the smooth distribution curve, of the density of states $N(W)$ for free electrons, becomes split up into allowed and forbidden bands, Figure 2.6(c) and (d). The parabolic shape is only retained for low energy, long wavelength electrons, which suffer no Bragg reflections, and so are not affected by the periodicity of the lattice.

We have so far been concerned with the band structure of the conduction, or unbound, electrons. Bands of energy levels also arise for *bound* electrons in the solid state. In a free atom, electrons occupy discrete or sharp energy levels. When two similar atoms are brought together, each two-electron state of the composite two-atom system becomes split into two energy levels, one higher and one lower than the corresponding energy of the free atom state. As the interatomic separation decreases, the splitting becomes larger and depends on the extent of the overlap between the electronic wave functions of the two atoms. The splitting is therefore greatest for the outermost atomic electrons, with extended wave functions, and least for the inner electrons. If more atoms are brought together, more discrete energy levels are created; until for a solid lattice of n atoms per unit volume, the n levels, formed from each free atom electron state, are so close together that they make an almost continuous band. The width of each band depends on the degree of overlap of the bound electron wave functions on adjacent atoms, and is largest for the outermost electrons (Figure 2.7). In fact, the

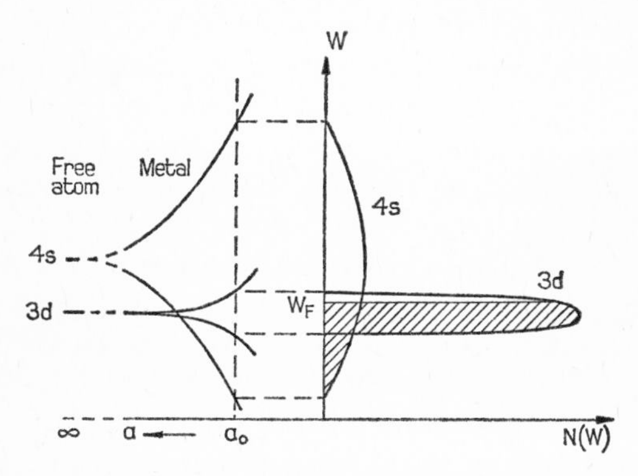

FIG. 2.7. To the left is shown the variation of the band widths of the 3*d* and 4*s* states as a function of interatomic spacing *a*. To the right is shown the density of states $N(W)$ for each band. The shaded portion corresponds to a 3*d* transition metal, with a large electronic specific heat

band is so wide for the valence electrons, that they lose their bound character and behave as unbound conduction electrons, in the manner outlined above.

Overlapping of the bands may, or may not occur. For example, in the 3*d* transition metals, 3*d* and 4*s* electron states have similar energies in the free atom, giving rise to transition element properties, like variable valency. In the solid state, the narrow 3*d* and broad 4*s* bands overlap, and can contain up to 10 and 2 electrons per atom, respectively. In Fe, Co and Ni, the 3*d* band is not completely filled, and there is a very large contribution to the density of states $N(W)_F$ at the Fermi surface, from the narrow but dense 3*d* band (Figure 2.7). This large value of $N(W)_F$ accounts for the abnormally high electronic specific heats of the transition elements.

2.11. Specific Heat Anomalies

Besides lattice and electronic specific heats, which increase monotonically with increasing temperature towards a limiting, constant (classical) value, there are many other specific heat phenomena. Among these, are excess specific heat bumps or anomalies, which are observed over a limited temperature range of one, or at the most two, orders of magnitude in temperature. These bumps are not really anomalous, except perhaps to the original observers who did not expect them, but the use of the term, *anomaly*, continues to the present day.

Specific heat anomalies may, in general, be divided into (a) Schottky and (b) co-operative, order-disorder, or λ-type processes. In both types, the internal energy and entropy vary continuously, and perhaps steeply with temperature, over the temperature range of the anomaly, but without any abrupt step-like variation. This behaviour is like that of a second-order thermodynamic transition, but only the λ-anomalies can be classified in this way. Neither type should be confused with first order thermodynamic transitions, where the internal energy, and entropy, change abruptly at a particular temperature, by amounts described in terms of a latent heat of transition.

2.12. The Schottky or non-Co-operative Anomaly

Atoms and molecules, in the solid state, frequently possess a small number of internal discrete quantised energy levels, quite separate from the spectrum of levels arising from the collective motion of the lattice. These internal states may arise in phenomena as diverse as molecular rotation, crystalline field interaction with paramagnetic ions, and nuclear hyperfine structure splittings.

Provided the atoms behave independently, in the sense that the thermal population of the energy levels is independent of the state of neighbouring atoms, then the atoms will be distributed among their internal levels according to simple Boltzmann statistics. The probability of occupying an energy state, with energy ΔE above the ground state, will be proportional to $\exp(-\Delta E/kT)$.

Consider a system of atoms, each with a singlet excited state separated by an energy ΔE from a singlet ground state. Then at $T = 0°K$, only the ground state will be occupied, while at $T \gg \Delta E/k$, equal numbers of atoms will populate the two states. As the temperature of the system is varied, a rapid change in the number distribution will take place over a relatively small temperature interval around T equal to $\Delta E/k$. The change in entropy and internal energy will be correspondingly large in this temperature range, and the associated excess specific heat will be an unsymmetrical bell-shaped curve (Figure 2.8(a)), reaching a maximum at $T = 0{\cdot}417\Delta E/k$, described by the relation

$$C_{\text{Schottky}} = R\left(\frac{\Delta E}{kT}\right)^2 \frac{\exp(\Delta E/kT)}{[1 + \exp(\Delta E/kT)]^2}$$

One important property, of a Schottky specific heat anomaly, is that for temperatures well above the specific heat maximum, the specific heat of the 'high temperature tail' varies very closely as $(\Delta E/kT)^2$. Thus, provided the excess specific heat can be satisfactorily separated from other contributions, the value of ΔE can be determined from the $1/T^2$ tail, at temperatures considerably greater than $\Delta E/k$. For example, the value of

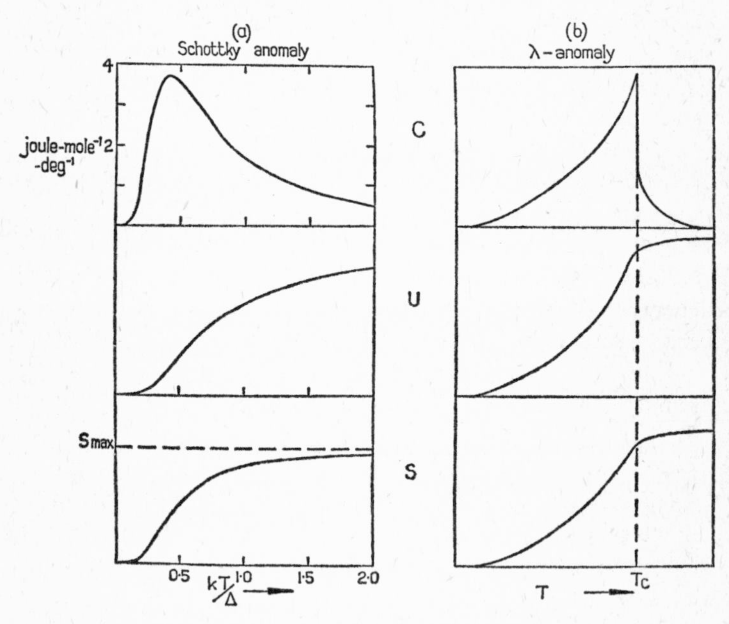

FIG. 2.8. The specific heat C, the internal energy U, and the entropy S plotted against T for

(a) a Schottky anomaly, associated with a singlet ground state and a singlet excited state separated by energy ΔE.

(b) a typical order-disorder λ-anomaly.

$\Delta E/k$, equal to 0·008°K, for the splitting between the nuclear hyperfine levels in antiferromagnetic α-manganese, was determined from specific heat measurements of the high temperature tail between 0·4°K and 1·2°K.

2.13. The co-operative or λ-anomaly

The Schottky anomaly is rather an idealised process, because atoms and their internal excitation levels do not behave independently, especially when they are closely packed in the solid state. Nevertheless, it can be described reasonably well mathematically, provided ΔE and kT are smaller or greater, by at least a factor of ten, than the co-operative energy of interaction ΔW between neighbouring atoms, ions or molecules.

This co-operative interaction tends to produce a spontaneous ordering, or entropy reduction, of a solid state system below a temperature $T = \Delta W/k$. The most common examples are the magnetic transitions from a disordered paramagnetic state to an ordered ferromagnetic or anti-ferromagnetic state. Other examples include the normal-superconducting transition in zero field, and order-disorder phenomena in alloys.

The transition from the low temperature ordered state to the high temperature disordered state is a cumulative process. As the temperature increases and approaches $\Delta W/k$, the probability of an atom occupying an excited state in thermal equilibrium depends, not only on the Boltzmann distribution law, but very strongly on the degree to which the neighbouring atoms already occupy that state. Thus every atom which passes into the excited state makes it easier for the next one to do so, and a cumulative process results.

The associated excess specific heat rises with increasing steepness as the temperature rises, and then falls abruptly to a small value when the transition to the disordered state is complete. The shape of the excess specific heat curve is like the Greek letter λ, and hence the transition is called a λ (lambda)-anomaly. The temperature at the specific heat peak usually defines the transition temperature T_c. Despite the abrupt nature of the specific heat, both entropy and internal energy increase smoothly through T_c from the ordered to disordered state, the increase being most rapid at T_c (Figure 2.8(b)). An example of both Schottky and λ specific heat anomalies occurring together in the same solid is given in Chapter 5, Figure 5.3(b).

3 Transport Effects in Dielectrics and Metals

3.1. Introduction

In the solid state, both heat and electrical conduction can be treated as arising from the flow of particles through the atomic lattice; electrons and phonons carrying heat energy along a temperature gradient; electrons and positive holes carrying charge along an electric potential gradient. A resistance to the particle flow arises from scattering processes, which can frequently be described, quite satisfactorily, in a way analogous to that used in the simple kinetic theory of a gas. We then talk in terms of a mean free path, l, between collisions, and a relaxation time, or average time between collisions, $\tau = l/v$, where v is the average particle velocity.

More rigorously, the probability $P(t)$ of a collision occurring within time t from the previous collision, is related to the relaxation time τ by

$$P(t) = 1 - e^{-t/\tau} \tag{3.1}$$

'$1/\tau$' may be determined formally, using time dependent perturbation theory, and will be an average of terms like

$$\frac{1}{\tau(k_1)} = |\int \psi(k_1) \,.\, V \,.\, \psi(k_2)\, dk_2|^2 \tag{3.2}$$

where $\tau(k_1)$ is the relaxation time from a particular initial state. $\psi(k_1)$ and $\psi(k_2)$ are wave functions describing the initial and scattered states, with associated wave vectors $\mathbf{k}_1$ and $\mathbf{k}_2$ respectively. V is the scattering perturbation in the potential energy of the particle, and the integral is taken over the available scattered states which the particle can occupy after scattering.

Since the particles (phonons and electrons) obey the laws of quantum statistics, there is a discrete spectrum of quantum states, only some of which may be empty for a particle to occupy after scattering. The density of these available empty scattered states, the energy exchanged in the scattering process, and the intensity of V will determine the probability of scattering. Fortunately, in many cases, this formalism reduces to the simple concept of a mean free path. The scattering processes can, then, be described by a simple kinetic theory treatment of billiard-ball-type collisions. We shall follow this simple treatment, bearing in mind that when the interpretation runs into difficulties, we should return to a more complete treatment using the formalism of equation (3.2.)

3.2. Heat Conduction Processes

All measurements of the thermal conductivity, K (watt cm^{-1} deg^{-1}), of a material depend on the simple fact that the heat transport is proportional to the temperature gradient along the specimen, i.e. the rate of heat flow $\dot{Q} = -\dfrac{KA\Delta T}{L}$ watts where A cm^2 is the cross-sectional area, and ΔT is the small temperature difference across a length L cm.

The two processes by which heat energy may be transported through a solid are:—

1. Conduction by the quantized lattice vibrations or phonons, which is best demonstrated in pure dielectric single crystals, like diamond or sapphire.

2. Conduction by the 'free' electrons in metallic solids, which is typified in measurements on pure metal single crystals.

Although both conduction processes can occur together, one is predominant in many cases. For example, in fairly pure metals, most of the heat is carried by the conduction electrons, and relatively little by the phonons. In a dielectric or insulating solid with no free electrons, the heat is carried entirely by the phonons.

Treating heat conduction as the flow of particles carrying heat energy down a temperature gradient, a simple treatment leads to a dimensional equation for the thermal conductivity (analogous to the thermal conduction equation, obtained in the simple kinetic theory of a gas).

$$K = \tfrac{1}{3}C_v v l \qquad (3.3)$$

C_v is the specific heat, at constant volume, of the heat carriers, v is their velocity, and l their mean free path between collisions, or scattering processes. The thermal conductivity, and its variation with temperature, is then determined by obtaining C_v, v and l separately, and combining them in equation (3.3.)

For each conduction process, phonon or electron, there are a number of mechanisms whereby the heat carriers are scattered. Each scattering mechanism restricts the heat flow and, therefore, contributes to the thermal resistance W (defined as the reciprocal of the thermal conductivity). To a good approximation, the various scattering mechanisms act independently, and the total thermal resistance is just the algebraic sum of the separate resistances, arising from each scattering mechanism.

i.e. $$W\,(\text{phonon}) = W_p = W_{pA} + W_{pB} + \ldots = 1/K_p$$

and $$W\,(\text{electron}) = W_e = W_{ea} + W_{eb} + \ldots = 1/K_e$$

where pA, pB, etc. indicate phonon scattering mechanisms, and ea, eb, etc. relate to electron scattering mechanisms. Now, phonon and electron heat conduction take place almost independently, and so the total thermal conductivity is very closely equal to the sum of the separate phonon and electron conductivities.

i.e.
$$K = K_p + K_e$$
$$= \frac{1}{W_p} + \frac{1}{W_e}$$

We shall now consider phonon and electron conduction separately, and outline the relative importance of the various scattering mechanisms in determining the total thermal conductivity. We shall confine our attention to two temperature ranges, in order to simplify discussion; low temperatures, where T is less than $\theta/20$ (or $\theta/10$), and high temperatures, around room temperature, where T is of the order of θ.

3.3. Heat Conduction by Phonons in a Dielectric Material

3.3(a). *Boundary Scattering.* ($T < \theta/20$). In general, as the temperature decreases, the thermal conductivity of a pure dielectric crystal increases to a maximum at a temperature of about $\theta/20$, in the range between 10°K and 70°K, with a value as high as 50 times the room temperature conductivity (Figure 3.1). At temperatures above the maximum, $T > \theta/20$, the conductivity is restricted by phonon-phonon scattering processes and decreases approximately as $1/T$, but the detailed variation in the region of the maximum is complex.

At temperatures below the conductivity maximum, $T < \theta/20$, the conductivity varies with T^3. In this temperature region, the phonons are only scattered when they reach the crystal boundaries. Only the low frequency, long wavelength ($\gtrsim$ 100Å) phonons are thermally excited, with a velocity of propagation equal to the velocity of sound v_0. The wavelength is too long for the phonons to 'see', and be scattered by, lattice discontinuities of atomic dimensions, and the phonon density is low, so that phonon-phonon scattering is negligible.

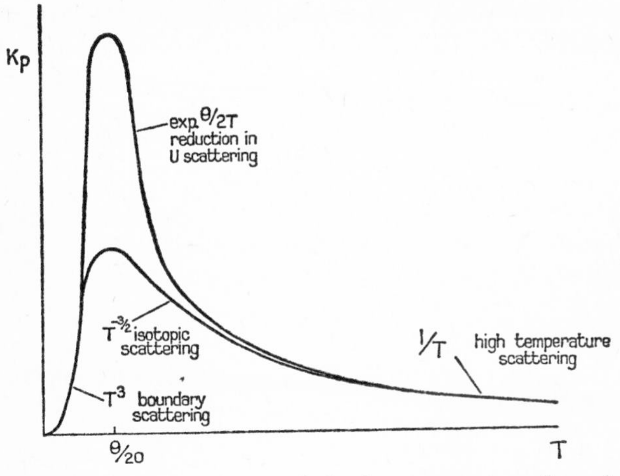

FIG. 3.1. General picture of the low temperature thermal conductivity of dielectric crystals, showing the steep exponential rise due to reduction of *U*-processes in isotopically pure crystals, and the less steep $T^{-3/2}$ rise in impure or mixed isotopic crystals.

In a rod-shaped, single crystal specimen, the phonons behave like a low pressure Knudsen gas passing along a tube, and are scattered only at the physical boundary of the crystal. The phonon flow depends on the diameter of the specimen, and also on the smoothness of the surface. The conductivity will be greatest when the phonons are specularly reflected from a polished surface, and will be reduced, as is experimentally observed, when the surface is roughened. The mean free path of the phonons is a constant, determined by the diameter d of the specimen. The specific heat of the phonons is just the lattice specific heat, proportional to T^3 in this temperature range. Substituting $C(\propto T^3)$, $v(= v_0)$ and $l(\propto d)$ in equation (3.3), the dimensional conductivity equation, we obtain

$$K_p \text{ (boundary)} = \text{constant} \times T^3 d$$
$$\text{or } W_p \text{ (boundary)} = \text{constant}/T^3 d$$

The conductivity is proportional to T^3, with the curious result that it depends on the dimensions of the crystal, being higher for a larger sample (Figure 3.2).

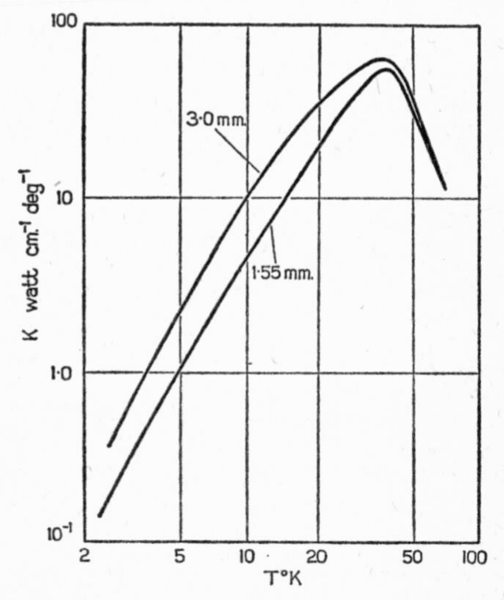

FIG. 3.2. Thermal conductivity of the same synthetic sapphire crystal, ground to successive diameters of 3·0 mm and 1·55 mm. At the lowest temperatures, the conductivity is limited by boundary scattering, and is proportional to the specimen diameter. (after Berman, Foster and Ziman, 1955)

For a polycrystalline dielectric, the phonons are scattered mainly at the boundaries of the crystallites, rather than at the physical boundary of the specimen. The mean free path is reduced to a small constant value, and the conductivity is correspondingly reduced in magnitude, with the T^3 variation maintained for $T < \theta/10$.

3.3(b). *Phonon-Phonon Scattering.* The potential energy of atoms in a crystal lattice has the form $Ax^2 + Bx^3 + Cx^4 + \ldots$, where x is the displacement from the equilibrium position. If B, C, etc. $= 0$, the motion of the atoms is purely harmonic, and from the principle of superposition of harmonic waves, no mechanism can exist for the interaction or collision between two phonons. If B, C, etc. are not zero, the motion is anharmonic, and readily provides a scattering mechanism between phonons. The most direct evidence that lattice forces *are* anharmonic, is provided by the fact that all solids expand when heated.

The theory of phonon-phonon scattering is a complex problem and was first developed in detail by Peierls (1929). At high temperatures ($T \gtrsim \theta$), the thermal conductivity is observed to be proportional to $1/T$ and can be explained quite simply. Since the lattice specific heat is constant, the lattice energy ($= \int C dT$), and hence the density of phonons, increases proportionally with T. The probability of any phonon-phonon scattering process is independent of the actual phonon-phonon interaction at these temperatures and depends only on the phonon density. The mean free path l (inversely proportional to the scattering probability) varies inversely with the density of phonons, and is therefore proportional to $1/T$. Combining these considerations of C_v and l, and taking the phonon group velocity to be constant, equation (3.3) reduces to

$$K_p\,(\text{phonon}) = \text{constant} \times 1/T$$

or

$$W_p\,(\text{phonon}) = \text{constant} \times T,$$

in agreement with experiment.

At lower temperatures, within the narrow range $\theta/20 < T < \theta/10$ (just above the conductivity maximum), K increases more rapidly than $1/T$ with decreasing temperature, and varies approximately as $\exp \theta/2T$. This variation has been observed, in the appropriate temperature ranges, for a number of isotopically pure, dielectric solids. Our simple picture of scattering processes depending only on the phonon density is inadequate at these temperatures, and we must examine the phonon-phonon interaction more closely. In general, we find that two types of collision can take place:—

1. Normal or *n*-processes, in which the total wave vector of the phonons is conserved (equivalent to the conservation of momentum). For example, the collision of two phonons may give rise to a third, according to the vector equation,

   $$\mathbf{k}_1 + \mathbf{k}_2 = \mathbf{k}_3$$

   Collisions of this type do not change, or provide a resistance to, the energy flow, but they are important for establishing an equilibrium distribution of phonons, at a given temperature.

2. *U*-(umklapp or flipping over) processes, in which momentum is not conserved, and the total wave vector, of two colliding phonons, changes by a vector quantity **g**, equal to 2π times the reciprocal lattice vector, according to the relation

   $$\mathbf{k}_1 + \mathbf{k}_2 = \mathbf{k}_3 + \mathbf{g}$$

It is not possible for the vector sum of the wave vectors of two phonons to be greater than $\mathbf{k}_m$, corresponding to the cut-off frequency ν_m of the lattice. Thus a transformation, using **g**, is required to allow the resulting phonon to have a meaningful resultant wave vector, smaller than $\mathbf{k}_m$. This transformation changes the direction of the energy flow (umklapp or flipping-over) and gives rise to a thermal resistance.

On average, each of the interacting phonons must have a wave vector greater than $\frac{1}{2}\mathbf{k}_m$, for a *U*-process to take place. In other words, the frequency of the phonons must be greater than $\frac{1}{2}\nu_m$, where $h\nu_m = k\theta$ according to the Debye theory of lattice specific heats.

The probability of exciting these phonons at low temperatures is proportional to:

$$\exp(-\tfrac{1}{2}h\nu_m/kT) \quad \text{or} \quad \exp(-\theta/2T)$$

Hence, the probability of a U-process occurring is proportional to $\exp(-\theta/2T)$, and leads to a thermal conductivity

$$K_p\,(\text{phonon}) = AT^n \exp(\theta/2T)$$

$$(\text{in the region } \theta/20 < T < \theta/10)$$

where A is characteristic of a crystal, and contains the anharmonicity of the lattice forces; n is a constant, but difficult to determine experimentally, because the exponential term varies much more rapidly than T^n.

At high temperatures, ($T \gtrsim \theta$), most of the phonons have wave vectors sufficiently large for U-processes to occur at each collision. Hence phonon-phonon scattering depends simply on the density of phonons, and leads to a $1/T$ variation of the conductivity (Figure 3.1).

3.3(c). *Phonon scattering by point defects and isotopic mixtures.* It is clear that imperfections in a regular crystal lattice will introduce additional scattering, and increase the thermal resistance. The extra resistance will depend on the type, density and distribution of these scattering centres and will vary over the phonon spectrum. Quantitative treatment is difficult to carry out satisfactorily. A good example of the reduction in conductivity by point defects, produced in quartz by successive thermal neutron irradiations, is shown in Figure 3.3. Note how the conductivity maximum is flattened with increasing concentration of point defects.

While the scattering due to a single atom with slightly different isotopic mass is small, the abundance of scattering centres in an isotopic mixture is very large. If a dielectric crystal is not almost isotopically pure, the phonon mean free path, due to isotopic scattering, is smaller than that for U-processes, in the region of the conductivity maximum, $\theta/20 < T < \theta/10$. Thus isotopic

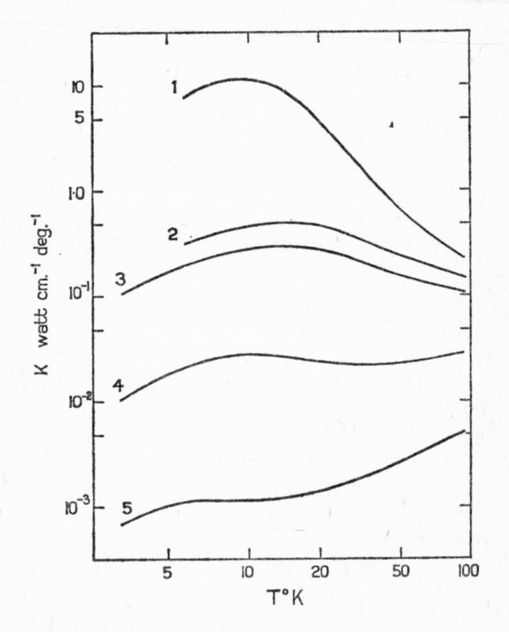

FIG. 3.3. Reduction in thermal conductivity by point defects, produced in a quartz crystal by successive thermal neutron irradiations. (1) quartz crystal, (2) after cumulative dose of $1{\cdot}8 \times 10^{18}$ neutron cm^{-2}, (3) $4{\cdot}3 \times 10^{18}$, (4) 34×10^{18}, (5) quartz glass. (from Berman, 1953)

scattering is the dominant process, and leads to a conductivity proportional to $T^{-3/2}$, which will mask the exponential variation expected for U-processes. The effect of changing the isotopic mixture in germanium (where phonon conduction is the dominant mechanism) is shown in Figure 3.4(a).

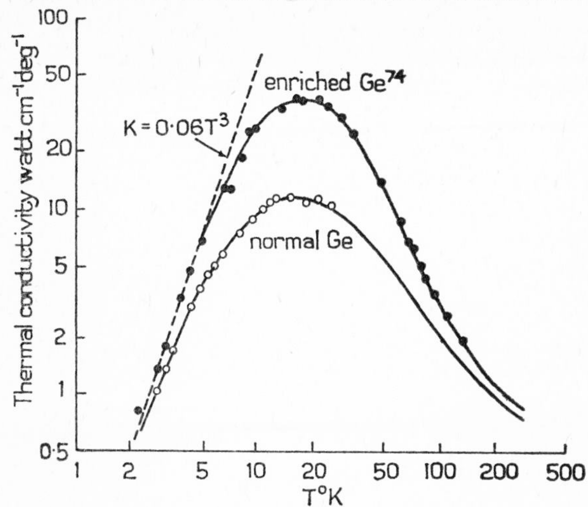

FIG. 3.4. Effect of isotopic scattering.

(a) The enriched sample of Ge^{74} has a higher conductivity than the natural sample. (from Geballe and Hull, 1958)

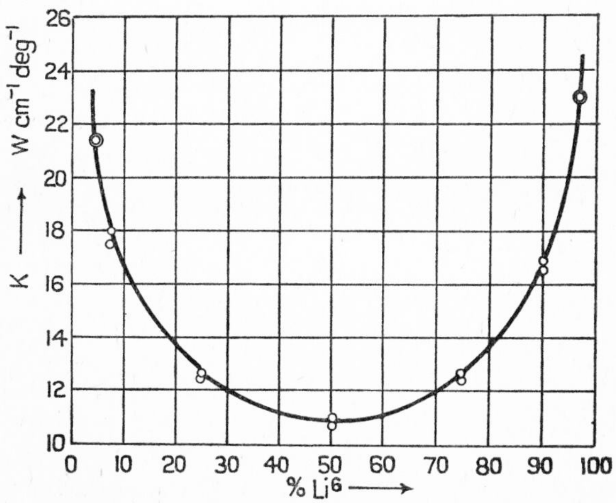

(b) Variation of conductivity with isotopic mixture of the system Li^6F—Li^7F at 30°K. (after Berman, 1959)

A convincing demonstration of the variation in conductivity at constant temperature of lithium fluoride, with different isotopic mixtures of Li^6 and Li^7, is shown in Figure 3.4(b).

3.3(d). *Thermal Conductivity of Glasses and Plastics.* In comparison with single dielectric crystals, the conductivities of commercial dielectric materials, like glasses, nylon, perspex, and other plastics, are very much smaller (about 10^{-3} times) at all temperatures. In general, the conductivity decreases continuously as the temperature is lowered, and no maximum occurs. For amorphous substances, like glass and plastics, there is no regular crystal lattice, and the mean free path of the phonons is limited to a small constant value, of the order of the interatomic spacing. The mean free path is also limited to a small constant value in microcrystalline materials like nylon. The conductivity is, therefore, very small, and proportional to the specific heat (see Figure 3.5). The small values of their

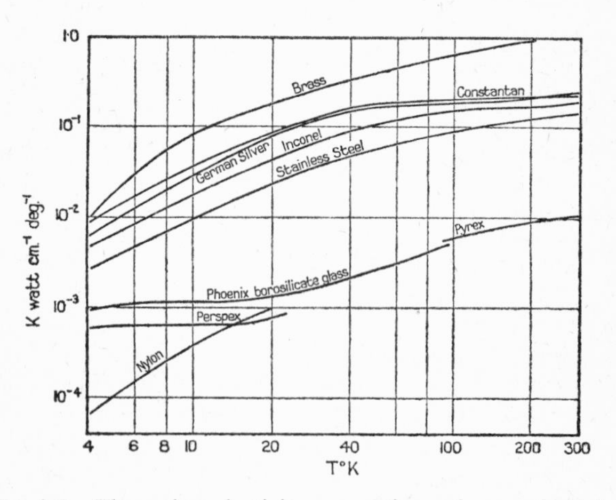

FIG. 3.5. Thermal conductivity curves of some low-conductivity commercial alloys and dielectric materials.

conductivities make these commercial materials ideally suitable for thermal insulation, at all temperatures down to the very lowest working temperatures ($< 10^{-2}$°K).

3.4. Heat Conduction in a Metal

3.4(a). *Importance of the Conduction Electrons.* In a metal, heat is transported mainly by the conduction electrons. Phonons are strongly scattered by the large number of 'free' electrons; the phonon mean free path is consequently small, and phonon heat conduction is suppressed. The reciprocal mechanism, namely the scattering of the conduction electrons by the phonons, together with impurity scattering, provide the limit to heat conduction by electrons.

The conductivity of a pure metal rises with decreasing temperature to a maximum at a temperature of about $\theta/20$, in a way somewhat similar to that of a pure dielectric. However, the variation with temperature is completely different on both sides of the maximum (Figure 3.6(a)). Impurity scattering is dominant below the maximum, while phonon scattering is dominant above the maximum.

3.4(b). *Impurity Scattering.* In the most pure metal crystal, conduction electrons can travel several hundred atomic spacings before being scattered. However, the presence of chemical impurities (foreign atoms), or physical impurities (lattice defects, dislocations, grain boundaries), provide static scattering centres, which limit the mean free path of the electrons to smaller constant values.

The important feature, of electron heat transport in a metal, is that we only need to consider electrons with energies close to the Fermi energy W_F. These are the only electrons which are allowed, by the Pauli exclusion principle, to be thermally excited into empty, higher, energy states, just above the Fermi

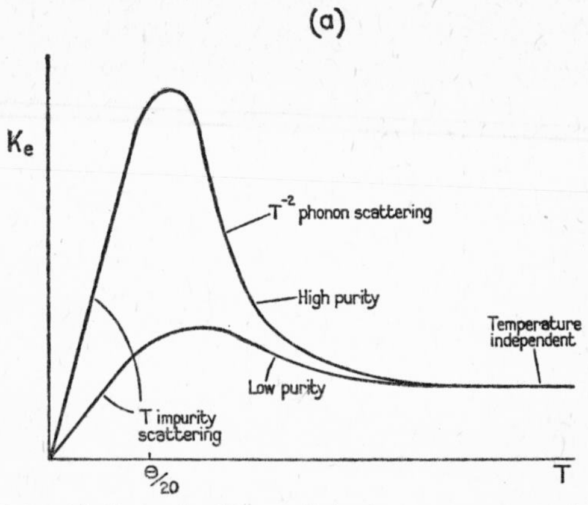

FIG. 3.6. (a) General picture of the low temperature thermal conductivity of high purity and low purity mteals.

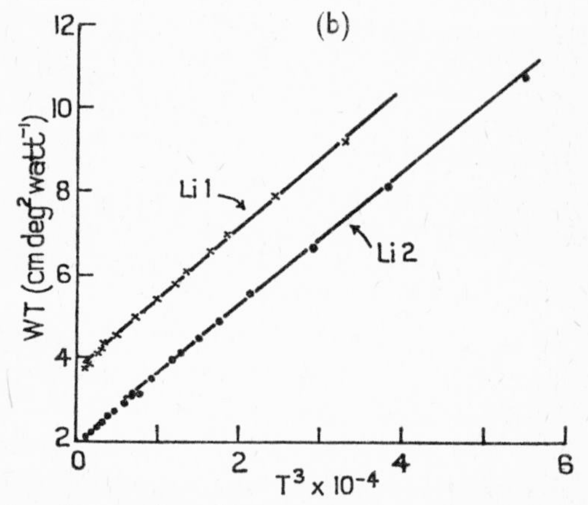

(b) Plot of WT against T^3 below 40°K for two lithium specimens of different purity. (from Rosenberg, 1956)

surface. The heat carriers, therefore, all have approximately the same kinetic energy, and consequently the same velocity, approximately independent of temperature.

The electronic specific heat is proportional to T, and since v and l are constant, the thermal resistance due to impurity scattering is given, using equation (3.3), by

$$W_e \text{ (impurity)} = \beta/T$$

β is a constant for a given sample of metal, being very sensitive to the chemical purity, and previous annealing treatment, of the sample.

3.4(c). *Phonon scattering of conduction electrons.* As the temperature rises above $\theta/30$, the scattering of electrons by the phonons becomes increasingly important, in comparison with the scattering by impurities. Up to a temperature of $\theta/10$, the lattice energy is proportional to T^4. The average energy of each phonon is kT, and so the density of phonons is proportional to T^3. The probability, of an electron-phonon scattering process taking place, depends directly on the phonon density, and so the mean free path, l, of the electrons (inversely proportional to this probability) varies as $1/T^3$.

Using equation (3.3), the thermal resistance due to phonon scattering is then:

$$W_e \text{ (phonon)} = \alpha T^2$$

where α is a constant, characteristic of the metal, and independent of the purity of the sample.

At temperatures above $\theta/10$ (i.e. above the T^3 specific heat region), W_e (phonon) continues to rise with increasing temperature, approaching a constant value at high temperatures. When $T \gtrsim \theta$, the phonon density increases proportionally with T (see section 3.3(b)), and l varies as $1/T$. Introducing this temperature dependence for l in equation (3.3), we find W_e (phonon), and hence the total thermal conductivity, is temperature independent, as observed in metals at room temperature.

3.4(d). *The total electronic thermal conductivity.* Adding the two thermal resistances, the total resistance and hence the total electronic thermal conductivity is given (for $T < \theta/10$) by

$$W_e = 1/K_e = W_e \text{ (phonon)} + W_e \text{ (impurity)}$$
$$= \alpha T^2 + \beta/T \qquad (3.5)$$

This relation gives a maximum in the variation of K_e, at a temperature of approximately $\theta/20$ for pure metals. The maximum can be very high—50 watt cm^{-1} deg^{-1}, or more, for very pure specimens—considerably greater than room temperature conductivity values ($\sim$ 4 watt cm^{-1} deg^{-1} for copper).

A check on the validity of equation (3.5), for temperatures less than $\theta/10$, is easily made by rewriting it as

$$W_e T = \alpha T^3 + \beta$$

and plotting $W_e T$ against T^3 (see Figure 3.6(b) for lithium metal). A straight line graph is obtained, where the intercept β varies with the purity of the sample, and the slope α depends only on the metal.

3.5. Phonon and Electron Heat Conduction in Metallic Alloys

In general, the presence of even 1% of impurity atoms, alloyed with a metal, can reduce the electron mean free path, by a factor of ten or more, below the value for the relatively pure metal. The value of β is correspondingly increased, and the conductivity maximum is flattened. With increasing percentage of impurity atoms, the electron conductivity becomes reduced to such an extent by impurity scattering, that the suppressed phonon-conduction becomes important. For example, Lomer (1958) has shown that the electron and phonon contributions are equal, when 5% zinc is alloyed with copper.

In commercial alloys, both electron and phonon heat conduction can be very small. Low conductivity alloys, of particular use for constructing low temperature cryostats, include stainless steels, Inconel, German silver (in tube form) and constantan wire. Some conductivity curves are shown in Figure 3.5.

3.6. Thermal Conductivity of Semi-Metals and Semi-Conductors

Elements, like bismuth and antimony, have very few conduction electrons, but still have metallic properties, and are called semi-metals. The number of electrons is not sufficient to produce much phonon scattering, and phonon conduction is not suppressed, as in a metal. Hence most of the heat is conducted by phonons. For example, in antimony, the dominance of phonon conduction is shown by a characteristic T^3 boundary-scattering dependence, below the conductivity maximum at 15°K.

The thermal conductivity of semi-conductors is somewhat similar. In high purity germanium and silicon, phonons are responsible for all the heat conduction, with a T^3 variation below the conductivity peak, and a $T^{-3/2}$ isotope scattering dependence above the peak (see Figure 3.4(a)). In general, the phonon conduction is reduced progressively by increased doping with impurities.

It appears that electrons and holes make very little direct contribution to the heat transport in a semi-conductor.

3.7. Electrical Conduction in Metals

3.7(a). *General Behaviour.* An outline of the theory of electrons in metals was given in sections 2.7 to 2.10. In general, the occurrence of band structure makes the theoretical treatment of electrical conduction much more difficult than the treatment of electron heat conduction. However, the basic features, of electrical conduction in metals, can be described by considering the conduction electrons as a Fermi gas.

Let us first go back to the classical model of an electron gas. In this model, the electrons are assumed to move in random directions through the lattice, with a classical Maxwell-Boltzmann distribution of energies, or velocities. On applying an electric field **E**, the electrons acquire a drift velocity, limited by scattering processes, which provides a current density **j**. Applying simple kinetic theory, the electrical conductivity σ of this classical model is given by

$$\sigma = \frac{\mathbf{j}}{\mathbf{E}} = \frac{ne^2\tau}{m} = \frac{ne^2}{m}\frac{l}{v} \tag{3.6}$$

where n is the number of electrons per unit volume; τ is the relaxation time, or average time between collisions; l is the mean free path and v the mean velocity of the electrons; e and m are the electronic charge and mass, respectively.

Replacing this classical model by that of a Fermi gas, only one modification to equation (3.6) is required. Since only electrons at the top of the Fermi distribution can be accelerated, and gain energy, in an electric field, the value of v is approximately constant, and equal to v_F, corresponding to W_F. 'n' still remains as the total number of conduction electrons—a peculiarity of the Fermi gas, readily explained in larger textbooks on solid state physics. The velocity v_F is about 10^8 cm. sec^{-1} for most metals; τ is about 10^{-14} seconds, and l is 10^{-6} cm. or about 100 atomic spacings at room temperature (l becomes much greater at lower temperatures). Thus the change in electrical conductivity with temperature depends primarily on changes in the mean free path l.

The two main types of electron scattering are the same as those which limit electron heat conduction in a metal, namely impurity and phonon scattering. However, the effect of these scattering mechanisms on the temperature variation of the electrical conductivity is completely different. To a good approximation, the scattering mechanisms behave independently, and their associated resistances may be added together (known as

Mattheissen's rule), in the same way as the various thermal resistance contributions were added in section 3.2.

It is worth pointing out that the mean free path of an electron should be infinite in a perfectly regular lattice of atoms, free from all types of impurity, and free of phonon excitations, i.e. at temperatures close to absolute zero. In this case, the electrical conductivity would be infinite. In practice, it is impossible to avoid some impurity in the lattice and the conductivity is never observed to become infinite in normal metals.

The occurrence of superconductivity in certain metals at low temperatures, where the electrical conductivity really does become infinite (and the electrical resistance identically zero) is a completely different phenomenon.

3.7(b). *Impurity Scattering.* As in electronic heat conduction, the presence of impurities provide static scattering centres which limit the mean free path of the electrons to a constant value, independent of temperature at $T < \theta/30$. From equation (3.6), the resistivity $\rho(= 1/\sigma)$ due to impurity scattering is, then, a constant, equal to the *residual resistivity* ρ_0. In a single crystal specimen, ρ_0 can be reduced by careful annealing treatment to remove the physical impurities, but a minimum is set by the unavoidable traces of foreign atoms present.

3.7(c). *Phonon Scattering.* In section 3.4(c), we describe a mean free path for electrons scattered by phonons, very simply in terms of the density of phonons. This picture is adequate enough to give the correct temperature variation of the thermal resistance, but it is insufficient to describe the electrical resistance arising from phonon scattering.

At low temperatures ($T \ll \theta$), the actual interaction between electron and phonon is a weak one, in the sense that a single electron-phonon collision produces only small angle scattering of the electron, and the energy exchanged is small. This can be seen by considering a single collision as a billiard-ball type collision and comparing the relative magnitudes of the electron and phonon momenta, p, or their respective k vectors

$$(\text{since } k = 2\pi/\lambda = 2\pi p/h).$$

For $T < \theta$, the dominant lattice wavelength is of the order $2a\theta/T$, and the corresponding value of k(phonon) is $\pi T/a\theta$, where 'a' is the lattice constant. When the first electron band is half-full, as in the alkali metals, the value of k(electron) at the Fermi surface is approximately π/a. The average angle ϕ through which an electron is scattered by a phonon is given approximately by

$$\phi \approx k(\text{phonon})/k(\text{electron})$$
$$\approx T/\theta$$

Clearly, when T is very much less than θ, the angle of scattering will be small, and an electron will have to suffer a number of successive collisions before any given accelerated motion, in an electric field, is erased. From a statistical treatment, the effective mean free path l_{eff} associated with the *drift* of conduction electrons in an *electric field* is $1/(1 - \cos\phi)$ times the average path l_0 between small angle scattering collisions;

i.e.
$$l_{eff} = l_0/(1 - \cos\phi)$$
$$\approx l_0 \times 2\frac{\theta^2}{T^2} \quad (3.7)$$

and increases very considerably with decreasing temperature.

Small angle scattering, by phonons, also occurs during the transport of heat by electrons along a temperature gradient. However in heat transport, there is no net flow of electrons. At any point in the metal, the electrons flowing *down* the temperature gradient will, on average, be *slightly hotter* than the equal number flowing in the opposite direction. The small quantity of energy exchanged in a single phonon scattering event can be quite sufficient to cool a 'hot' electron below the average. Thus *each encounter with a phonon* is an effective random scattering process, and the mean free path of electrons during the transport of heat is l_0, inversely proportional to the density of phonons.

For $T < \theta/10$, the phonon density is proportional to T^3/θ^4; l_0 is therefore proportional to θ^4/T^3.

Hence $l_{\text{eff}} \propto \dfrac{\theta^6}{T^5}$, and from equation (3.6)

$$\rho(\text{phonon}) = \frac{T^5}{\theta^6} \times \text{constant (for } T < \theta/10) \qquad (3.8)$$

As the temperature increases, the average scattering angle, T/θ, becomes larger. When $T \gtrsim \theta$, each electron-phonon collision is a truly random process, and the mean free path depends only on the phonon density ($\propto T$ in this temperature region).

Hence l_{eff} varies as $1/T$ and

$$\rho(\text{phonon}) = T \times \text{constant (for } T \gtrsim \theta) \qquad (3.9)$$

A theoretical expression, which closely describes the variation of ρ(phonon) over the whole temperature range, is the Bloch-Gruneisen relation. This assumes a Debye phonon spectrum and gives

$$\rho(\text{phonon}) = \frac{BT^5}{\theta^6} \int_0^{\theta/T} \frac{z^5 dz}{(e^z - 1)(1 - e^z)}$$

where B and θ are chosen to give the best fit with experimental results.

For $T < \theta/10$, the relation reduces to

$$\rho(\text{phonon}) = 125 \frac{BT^5}{\theta^6} \text{ similar to equation (3.8)}$$

For $T \gtrsim \theta$, the relation becomes

$$\rho(\text{phonon}) = \frac{BT}{4\theta^2} \text{ similar to equation (3.9)}$$

3.7(d). *The total electrical resistivity.* Combining the two contributions, the total electrical resistivity is given by

$$\rho = \rho_0 + \rho(\text{phonon})$$

The behaviour of ρ at low temperatures is illustrated in Figure 3.7(a) by the results on two samples of sodium, of different purity. The figure shows how the two resistivity contributions may be separated. Below 4·2°K, the resistivity is almost completely determined by the residual resistivity ρ_0.

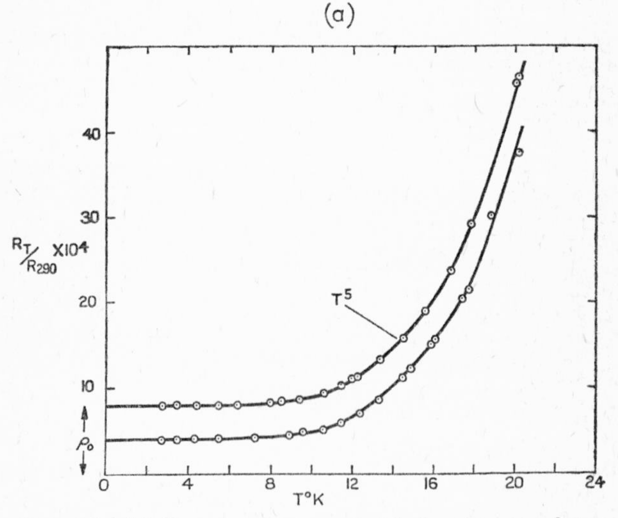

FIG. 3.7. (a) Low temperature electrical resistance of two sodium specimens of different purity (after MacDonald and Mendelssohn, 1950)

At higher temperatures, the variation of ρ(phonon) fits the Bloch-Gruneisen relation quite closely (Figure 3.7(b)). The approach to a linear variation of ρ with T, at temperatures above $\theta/5$, enables convenient temperature measurement to be made with metallic resistance thermometers.

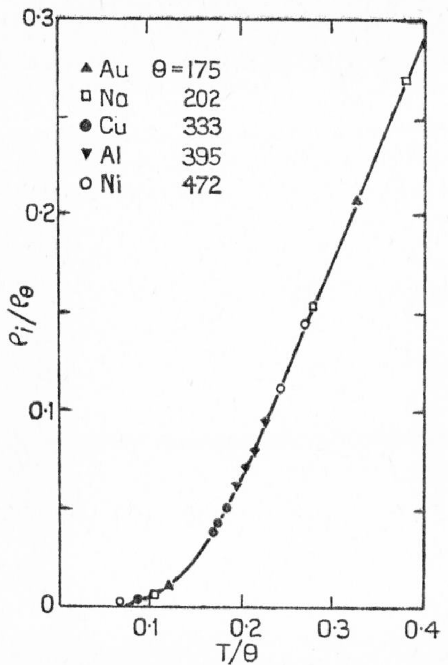

(b) A plot of the Bloch-Gruneisen relation for the electrical resistivity, together with experimental data for five metals (after Meissner)

3.8. Thermal and Electrical Conduction in 3d Transition Metals

We have already seen, in section 2.10, how the large electronic specific heat of $3d$ transition metals arises from the overlap of the $3d$ band with the $4s$ conduction band. Both bands are partially filled, and the main contribution to the electronic specific heat, γT, is provided by the high density of electron states at the top of the filled portion of the narrow $3d$ band. However, in spite of the high density of states, and consequently large number of heat and charge carriers available from the $3d$ band, the thermal and electrical conductivities of the $3d$ transition metals are small in comparison with those of the monovalent metals.

A qualitative explanation can be made by referring to the formal expression for the relaxation time τ in equation (3.2). $1/\tau$ is large when there are a large number of available scattered states, i.e. when the density of states at the Fermi surface is large. Consequently, τ is considerably shorter for $3d$ band electrons, than for $4s$ conduction electrons, with the result that $3d$ electrons contribute very little to thermal and electrical conduction processes.

Furthermore, since the $4s$ and $3d$ bands overlap, it is more easy for a $4s$ electron to be scattered into a $3d$ state, than into another $4s$ state. In other words, the $3d$ band tends to act as a non-conducting sink into which the heat and charge carriers are scattered and absorbed. This mechanism creates a large, and perhaps dominant, resistance, reducing both thermal and electrical conduction by $4s$ electrons.

A similar resistive mechanism can be expected to occur in other metals when two unfilled bands overlap at the Fermi surface. Theoretical treatment is difficult and usually incomplete.

3.9. The Wiedemann-Franz Law

Both heat and electrical conduction in metals arise from the motion of electrons with energies close to the Fermi energy W_F. The motion is restricted by similar scattering mechanisms, and we expect the two conductivities to be related, mathematically. Combining the kinetic expressions for K (equation (3.3)) and σ (equation (3.6)) to form the quantity $K/\sigma T$, we obtain

$$\frac{K}{\sigma T} = \frac{mv^2 C_e}{3ne^2 T} \cdot \frac{l(K)}{l(\sigma)}$$

where $l(K)$ and $l(\sigma)$ are the electron mean free paths applicable to heat conduction and electrical conduction respectively. For a Fermi gas, C_e is given from equation (2.17) by

$$C_e = \frac{n\pi^2 k^2 T}{2W_F}$$

$$\text{Hence } \frac{K}{\sigma T} = \frac{\pi^2 k^2}{3e^2}\left(\frac{mv^2}{2W_F}\right)\cdot\frac{l(K)}{l(\sigma)}$$

But mv^2 is just twice the kinetic energy of electrons at the Fermi surface, $2W_F$, and so the expression reduces to

$$\frac{K}{\sigma T}\left(\text{or } \frac{\rho}{WT}\right) = \frac{\pi^2 k^2}{3e^2}\cdot\frac{l(K)}{l(\sigma)}$$

$$= 2{\cdot}45 \times 10^{-8} \text{ when } l(K) = l(\sigma)$$

$$= L_0, \text{ the Lorentz constant,}$$

where K is expressed in watt cm^{-1} deg^{-1} and σ in ohm^{-1} cm^{-1}.

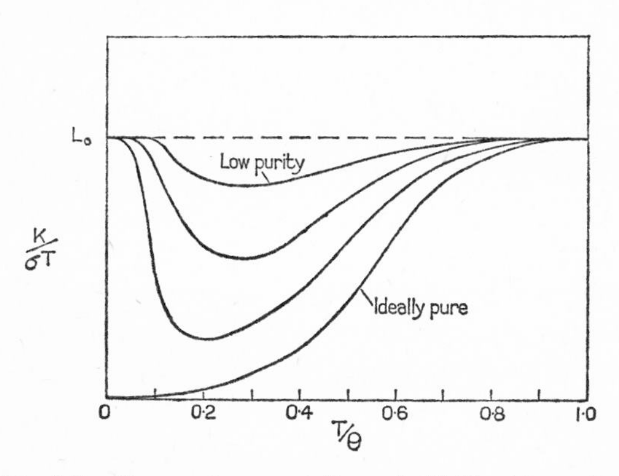

FIG. 3.8. Temperature variation of $K/\sigma T$ with purity. L_0 is the limiting value at low and high temperatures.

This relation, commonly called the Wiedemann-Franz law, can be derived much more generally, and can be shown to be valid even when the conduction electrons in a metal do not approximate to a Fermi gas model. The Lorentz constant, L_0, is a universal constant applicable to all metals, provided the condition, $l(K) = l(\sigma)$, is met.

The two mean free paths are identical at low temperatures, $T < \theta/10$, where impurity scattering is dominant, and also at high temperatures, $T \approx \theta$, when phonon scattering depends only on the phonon density. At intermediate temperatures, where low angle scattering by phonons is the dominant contribution to the electrical resistivity, $l(\sigma)$ is greater than $l(K)$ and the law does not hold. The general variation of $K/\sigma T$ is shown in Figure 3.8, where it can be seen that the more impure the metal, the closer does the law hold over the whole temperature range.

4 Superconductivity

4.1. Introduction

The discovery by Kamerlingh Onnes in 1911, that the electrical resistivity of mercury fell abruptly to zero below a temperature of 4·16°K, opened up another field of interest at low temperatures. A considerable number of metallic materials—elements, compounds and alloys—have been found to exhibit the same property of zero resistance, or of superconductivity, below a transition temperature T_c, characteristic of each material. Careful experiments have shown that, below T_c, the electrical resistance is identically zero, and the electrical conductivity is infinite. No Joule, or dissipative, heating accompanies the passage of an electric current, and a current flowing in a closed loop of superconducting material will persist indefinitely.

Superconductors may be broadly classified into Type I and Type II superconductors, with quite different behaviours in a magnetic field (see later). Type I superconductors include most of the pure superconducting elements, and many alloys, while Type II superconductors are usually alloys and intermetallic compounds.

Some of the Type II alloys and compounds have particular technological importance in their ability to remain superconducting in high magnetic fields ($\sim$100 kOe) and/or when carrying high current densities ($\sim 10^5$ amps cm^{-2}).

4.2. Properties of a Superconductor

4.2(a). *Occurrence.* So far, 25 metallic elements have been shown to have superconducting properties (see Table 4.1).

The transition temperatures range from 0·14°K for Iridium to 11·2°K for Technetium—the highest transition temperature is about 18°K for the compound Nb_3Sn. Although a considerable number of metals have been investigated below 0·1°K, no transition points have been found in this temperature region.

TABLE 4.1.

At. No.	Element	T_c°K	H_0(Oe)	At. No.	Element	T_c°K	H_0(Oe)
13	Al	1·197	105				
22	Ti	0·39		50	Sn	3·74	310
23	V	4·89	1200	57	La	~4·8	
30	Zn	0·93	55	72	Hf	0·37	
31	Ga	1·10	50	73	Ta	4·38	960
40	Zr	0·55		75	Re	1·70	
41	Nb	8·9	2600	76	Os	0·71	65
42	Mo	~0·95		77	Ir	0·14	
43	Tc	11·2		80	Hg	4·16	420
44	Ru	0·47	45	81	Tl	2·39	170
45	Rh	0·9		82	Pb	7·22	810
48	Cd	0·56	28	90	Th	1·37	
49	In	3·4	270	92	U	1·1	

Transition temperatures T_c, and critical fields H_0 (at $T = 0$°K), of the superconducting elements.

Notable exceptions from the list include the alkali metals, the noble metals, and the magnetic iron-group and rare-earth metals. Generally, it appears that antiferromagnetism and ferromagnetism are incompatible with superconductivity, and the presence of a magnetic impurity tends to suppress the occurrence of superconductivity.

All superconductors appear to have intermediate atomic volumes. For example, the compound Au_2Bi is a superconductor, while the pure elements Au and Bi, with respectively small and large atomic volumes, are not.

The importance of crystal structure is shown by the fact that white tin (tetragonal) is superconducting below 3·7°K, while grey tin (cubic) is not superconducting at all. Although superconductivity is not confined to any particular crystal structure, the importance of the part played by the lattice is demonstrated by the *isotope effect*. Precise measurements on different isotopes of the same metal show that $T_c\sqrt{M}$ = constant, where M is the isotopic mass. Since the frequency spectrum of lattice vibrations is expected to change slightly for different values of M, an important clue is provided towards the present-day concept of superconductivity as a collective interaction, between non-localised electrons and lattice vibrations.

An empirical relation has also been found between T_c and the average number of valency electrons in elements, compounds and alloys. Matthias demonstrated that there are pronounced maxima in T_c when the average number is 3, 5 and 7, and that this relationship can be used to predict the T_c of new superconducting materials.

4.2(b). *The superconducting—normal transition.* The transition from the superconducting state to the normal state in zero magnetic field, as indicated by the recovery of the full normal resistance, takes place over a finite temperature interval, which depends on the physical and chemical impurity content. A transition interval as small as 0·001°K has been observed in a carefully annealed, high purity, single crystal, but in impure specimens which have been highly strained, the transition may extend over 0·1°K or more (see Fig. 4·1(a)). T_c is defined as the temperature at which the recovery of the normal resistance commences.

The superconducting state is destroyed by the application of a magnetic field greater than a critical value H_c; the metal then

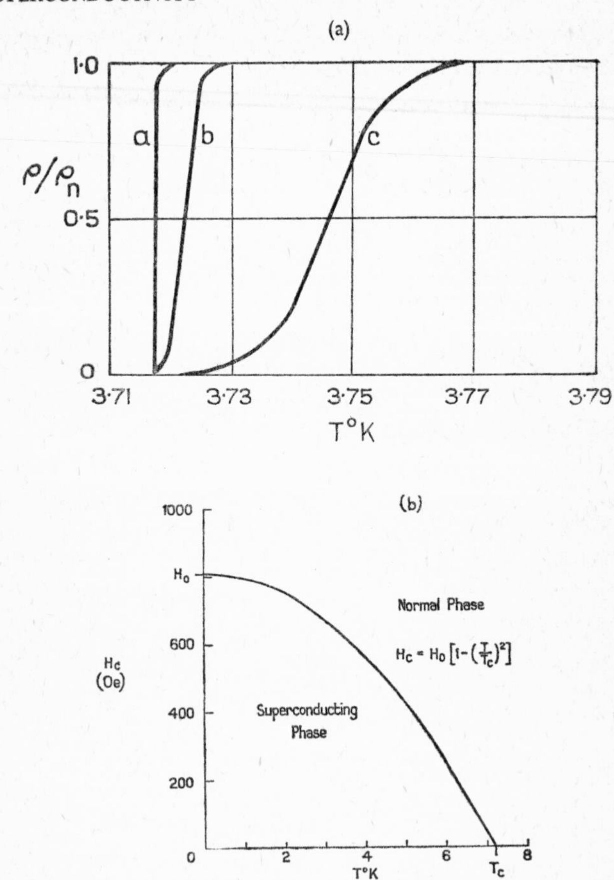

FIG. 4.1. (a) The superconducting transition of tin for (a) high-purity single crystal, (b) high purity polycrystal and (c) impure polycrystal. ρ_n is the normal state resistance. (from de Haas and Voogd, 1931)

(b) Variation of the critical field H_c with temperature, for superconducting lead.

returns to the normal state with a finite electrical resistance. H_c depends on temperature according to an approximate parabolic relation

$$H_c = H_0[1 - (T/T_c)^2]$$

Thus H_c is zero at T_c, and rises with decreasing temperature, approaching a constant value H_0 as T approaches 0°K (Figure 4.1(b)). Again, the sharpness of the recovery of the normal electrical resistance, as the field is increased through H_c, depends on the purity of the sample.

Of greater importance, the transition interval depends on geometry. For example, if the field is applied parallel to the axis of a long cylindrical specimen, the transition is sharp; if the field is perpendicular to the long axis, the recovery of the normal resistance is more gradual, commencing at about $\frac{1}{2}H_c$ for Type I superconductors. It is usual practice therefore, to measure H_c when the field is applied parallel to the axis of a long wire specimen.

For pure metals, and other Type I superconductors, H_0 varies from a few tens of oersted for metals with low T_c, up to a few kilo-oersted for high-purity niobium, $T_c \sim 9°K$ (Table 4.1). For Type II superconductors, the values of H_0 are generally higher, and may be as high as a few hundred kilo-oersted. However, the values of H_0 serve only as a secondary distinction between the two types.

The superconducting state may also be destroyed by passing a sufficiently large electric current. The critical current is defined simply as that which produces the critical magnetic field H_c at the surface of the conductor, and starts to restore the normal resistance. In a wire of radius, r, the critical current i_c is given by $i_c = \frac{1}{2}rH_c$. The full normal resistance may not be attained until a current of about $2i_c$ is reached.

4.2(c). *The Meissner effect in Ideal Types I and II Superconductors.* For ideal Type I superconductivity, the magnetic induction *B remains zero* inside the body of a superconductor, irrespective of the previous magnetic treatment of the sample. External magnetic flux cannot penetrate the body, which therefore appears to behave as a perfect diamagnetic material.

Using the relation between magnetic induction B, magnetic field H, and magnetic moment per unit volume M, namely $B = H + 4\pi M$, it can be seen that, since B is zero, $M = -H/4\pi$, the magnetic moment of a perfect diamagnetic material.

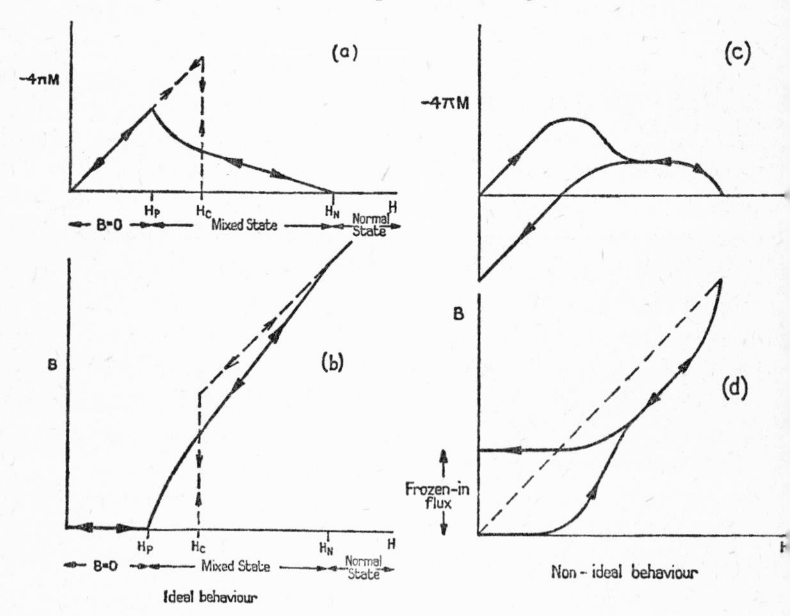

FIG. 4.2. (a) Magnetisation M of ideal Type I (broken line) and Type II (full line) superconductors.
(b) Magnetic induction B in ideal Type I (broken line) and Type II (full line) superconductors.
(c) Typical irreversible magnetisation of a non-ideal superconductor.
(d) Associated magnetic induction in a non-ideal super-conductor, showing frozen-in flux.

If a field $H(< H_c)$ is applied in the normal state, and the temperature is then reduced below T_c, all the magnetic flux is expelled from the body of the metal during the normal-superconducting transition. This expulsion of magnetic flux is called the *Meissner effect*.

The exclusion of magnetic flux arises from shielding supercurrents, induced in the surface by the external field, which create an opposing magnetic field just sufficient to reduce the resultant magnetic induction, within the body of the superconductor, to zero.

Plotting B or M against H, where the field H is parallel to the axis of a wire specimen, the difference in behaviour of ideal Type I and Type II superconductors becomes apparent (see Figures 4.2(a) and 4.2(b)). In both cases, when H is reduced to zero, B (or M) reduces to zero along a reversible path.

In Type I superconductors, flux penetration (or expulsion) takes place only when H_c is reached. With increasing field applied to a Type II superconductor, flux penetration commences at H_p, but the full normal resistive state is not restored until a very much higher field H_n (10–100 times H_p) is reached. Between H_p and H_n, the superconductor is said to be in the *mixed state*, when it breaks up into a finite number of laminae, of alternate superconducting and normal states, parallel to the applied field (Figure 4.3).

It should be noted that a similar mixed or *intermediate state* is created in Type I superconductors, when the magnetic field is perpendicular to the wire axis, with values between about $\frac{1}{2}H_c$ and H_c.

The interface, between normal and superconducting phases in the mixed state, has a surface energy of formation, α, the sign of which determines Type I or Type II behaviour. If α is positive, the total area of interface formed is limited by energy considerations; only a small number of laminae will form spontaneously and Type I behaviour occurs.

If α is negative, however, there are no energy limitations on the interface area, and a large number of extremely thin laminae or filaments can form, allowing the flux to penetrate almost homogeneously throughout the superconductor between H_p and H_n, the lower and upper critical fields respectively, giving rise to Type II behaviour. The importance of this spontaneous

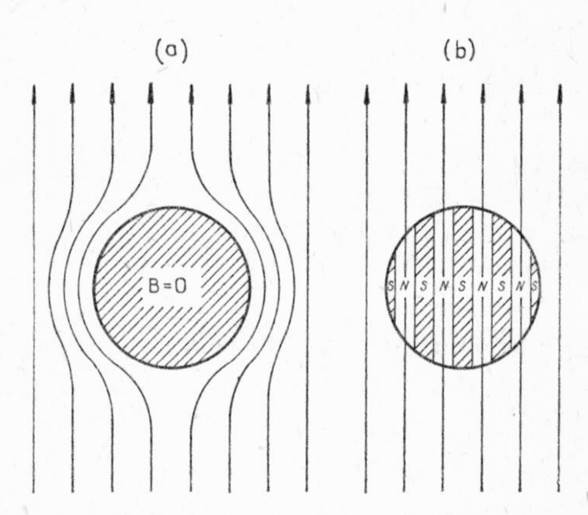

FIG. 4.3. Effect of magnetic field

(a) Flux expelled, $B=0$.

(b) Mixed state of alternate normal and superconducting laminae, parallel to the field.

formation of thin superconducting laminae lies with the fact, that the critical field of the laminae is very much greater than that for the bulk superconductor.

It was once thought that small inhomogeneities, strains and other impurities were necessary to make α negative and induce Type II behaviour. This is now believed not to be so; the impurities are only necessary to stabilise the superconducting filaments against attractive, coalescing forces, when they are carrying large supercurrents.

4.2(d). *Non-ideal behaviour of Types I and II Superconductors.* Unless great care is taken to use pure, well-annealed specimens, the ideal reversible behaviour in a magnetic field (as shown in Figures 4.2(a) and 4.2(b)), is frequently not observed. In physically and chemically impure specimens, the magnetic flux remains 'frozen-in' after applying and removing a field greater than H_c, and an incomplete Meissner effect may be observed. The non-homogeneous mixed state of alternate superconducting and normal phases is retained. It is believed that the superconducting laminae, or filaments, join up to form a three dimensional lattice, or 'sponge', enclosing regions of normal phase. On removing the field, induced super-currents remain, circulating the normal phase regions, giving rise to the frozen-in flux. Recent experiments have shown that the flux is quantised, from the observation that 'flux jumps' of one, or more, quantum units of $\frac{hc}{2e}$ ($= 2{\cdot}07 \times 10^{-7}$ gauss cm^{-2}) take place, as the applied field is slowly varied.

A whole range of non-ideal, or hysteresis, behaviours may be observed; see for example Figures 4.2(c) and 4.2(d).

4.3. The Two-Fluid Model of Superconductivity

On a simple model, the conducting fluid, or conduction electrons, may be divided into super-electrons, in some unspecified completely ordered state, and normal electrons, which are not in an ordered state. At finite temperatures, a fraction x of the electrons are super-electrons, the remaining fraction, $1 - x$, being normal electrons. x is a function of temperature, shown from specific heat measurements to vary in the form

$$x = 1 - (T/T_c)^4$$

Since super-electrons pass through the lattice with zero resistance, they carry all the current flowing, effectively short-circuiting the fraction of normal electrons, and the superconductor shows zero electrical resistance.

4.4. Electromagnetism of Superconductors, and Penetration Effects

Confining our attention to Type I superconductors, it is clear that, when a magnetic field is applied, the induced shielding super-currents must flow within some thin but finite layer at the surface, to maintain zero induction within the bulk of a superconductor. Within this surface layer, the magnetic induction B is not zero, and the applied field penetrates a short distance before falling to zero. It is also clear that a current flowing in a superconducting wire is confined to the same surface layer, in a manner analogous to the surface flow of high frequency alternating currents in normal metal conductors.

F. and H. London have shown that the normal Maxwell equations of electromagnetism have to be supplemented by two further equations for a superconductor, namely

$$\text{curl}\, \Lambda \mathbf{j}_s = -\mathbf{H}/c$$

$$\text{and} \quad \frac{\partial}{\partial t}(\Lambda \mathbf{j}_s) = \mathbf{E}$$

where Λ is a constant, characteristic of the material. The solution of these equations, for the condition when $\mathbf{H}$ and $\mathbf{j}_s$ do not vary with time (i.e. when $\mathbf{E} = 0$), are given by

$$\nabla^2 \mathbf{H} = \mathbf{H}/\lambda^2$$

and

$$\nabla^2 \mathbf{j}_s = \mathbf{j}_s/\lambda^2$$

where

$$\lambda = \left(\frac{\Lambda c^2}{4\pi}\right)^{\frac{1}{2}}$$

For the simple case of a semi-infinite superconductor, bounded by the $x = 0$ plane, an external field H_0, parallel to the y axis, falls off inside the superconductor according to the exponential relation

$$H_y = H_0 \exp(-x/\lambda)$$

The field falls by a factor $1/e$ at a depth λ, called the penetration depth, whose magnitude has been shown experimentally to be about 5×10^{-6} cm.

The induced super-current density j_s is in the z-direction, and also falls off within the superconductor with the same exponential behaviour.

4.5. The High Critical Field in Thin Superconducting Laminae

The H_c–T curve can be treated as an equilibrium line in a phase diagram separating superconducting and normal phases. If F_s is the molar Gibbs free energy of the superconducting phase, the Gibbs free energy in a field H is $F_s - \frac{1}{2}HMV$, V being the molar volume, and M the induced magnetisation, equal to $-H/4\pi$ for a perfect diamagnetic.

At the equilibrium curve ($H = H_c$), the Gibbs free energy of the two phases are equal, and hence

$$\Delta F_{\text{bulk}} = F_n - F_s = (H_c^2/8\pi)V \tag{4.1}$$

Within the penetration depth, the magnitude of the magnetisation M_d is smaller than that in the bulk material.

$$\text{i.e. } M_d = -\gamma H/4\pi \text{ where } 0 < \gamma < 1$$

and $$\Delta F_d = (F_n - F_s)_d = (\gamma H^2/8\pi)V$$

Clearly, ΔF_d only reaches the value of ΔF_{bulk} in a higher field. In other words, the penetration depth has a critical field $H_d = H_c/\sqrt{\gamma}$, higher than the bulk specimen. The effect is negligible for large specimens, where the surface volume is small in comparison with the bulk. However, in thin films and wires, with thickness of the order of λ, a considerable increase in critical field is observed.

At constant temperature, experiments show that for a thin film of thickness t cm,

$$H_d = H_c(1 + k\lambda/t) \tag{4.2}$$

and for a wire, radius r cm,

$$H_d = H_c(1 + k\lambda/r) \tag{4.3}$$

where k is of the order of unity, and is temperature dependent. If t, or r, is about $\lambda/10$, then it is clear that H_d can exceed H_c by a factor of about 10.

Let us now return to Type I and II superconductors in the mixed state. It has been observed directly, by experiment, that the laminae of a Type I superconductor are about 10^{-2} cm or $2000 \times \lambda$ thick. Hence, from equation (4.2), the difference between H_d and H_c is negligible.

In Type II superconductors, the thickness of each superconducting lamina or filament maybe smaller than 5×10^{-7} cm, or $\lambda/10$, with a critical field H_d greater than $10 \times H_c$. Thus a field H_d, equivalent to the upper critical field H_n in Figures 4.2(a) and 4.2(b), is required to drive a Type II superconductor completely normal. The lower critical field H_p, above which the mixed state sets in, is of the order of H_c.

On a scale large compared with λ (e.g. for a wire, 10^{-2} cm. diameter) the magnetic flux appears to penetrate almost homogeneously through a sample of Type II superconductor in the mixed state, and a supercurrent appears to flow uniformly through the whole cross-section.

4.6. Some Thermodynamics of a Bulk Type I Superconductor

Provided a complete Meissner effect is shown and no intermediate state exists, we can apply thermodynamics very simply to a bulk Type I superconductor.

From equation 4.1 $\quad F_n - F_s = (H_c^2/8\pi)V$

Since the entropy $\quad S = -\left(\frac{\partial F}{\partial T}\right)_H$

$$S_n - S_s = \Delta S = -\frac{H_c}{4\pi}\left(\frac{dH_c}{dT}\right)V$$

Differentiating the $H_c - T$ curve, we see that dH_c/dT is negative, approaching zero at $T = 0$.

Then ΔS has zero values at $T = 0$ (in agreement with the third law of thermodynamics), and at $T = T_c$ (when $H_c = 0$). At all

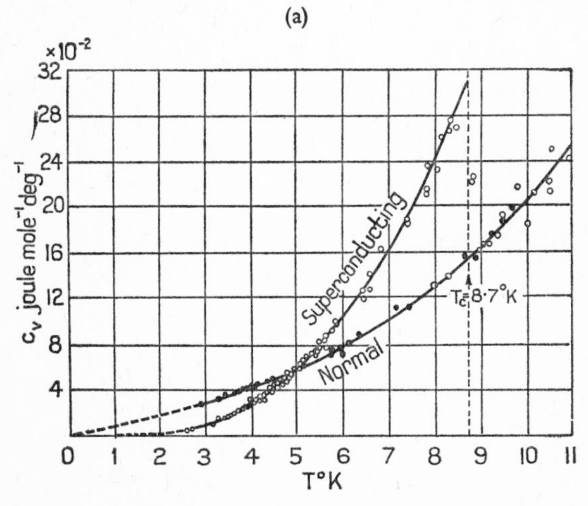

FIG. 4.4. (a) Specific heat of niobium in the normal and superconducting states, showing λ-anomaly at T_c (Brown, Zemansky and Boorse, 1953)

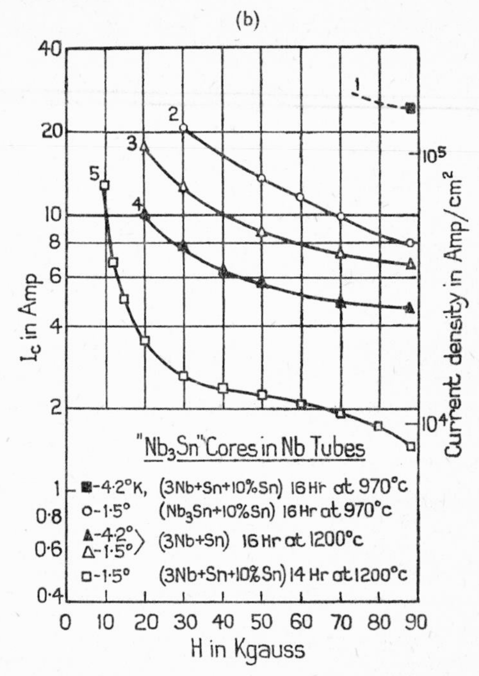

FIG. 4.4. (b) Variation with magnetic field of the critical current (and critical current density) in sintered Nb_3Sn wires. (from Kunzler, Buehler, Hsu and Wernick, 1961)

intermediate temperatures, ΔS is positive, showing that the superconducting state is more ordered than the normal state. Furthermore, the s–n transition at constant temperature T through the critical field, is a first order transition, with a latent heat $Q = T\Delta S$ reducing to zero at the critical temperature T_c and at $T = 0$.

The specific heat difference $C_s - C_n$ is given by

$$C_s - C_n = T\left(\frac{\partial S}{\partial T}\right)_H = \frac{T}{4\pi}\left[H_c\frac{d^2H_c}{dT^2} + \left(\frac{dH_c}{dT}\right)^2\right]V$$

in good agreement with specific heat measurements on the normal and superconducting phases. At the lowest temperatures, the specific heat difference is negative, becoming positive as T approaches T_c. The behaviour of niobium is shown as an example in Figure 4.4(a). Note the λ-type specific heat anomaly at T_c, characteristic of an order-disorder transition.

4.7. High-Field Superconductors and 'Super-Magnets'

The attainment of steady magnetic fields greater than 25 kOe over large volumes, with normal high current solenoids requiring the order of 1,000 kw. of D.C. power, is a considerable undertaking, needing large D.C. generators and complex cooling systems. Until recently, the idea of a superconducting solenoid capable of providing magnetic fields of even 10 kOe, with the enormous attraction of *no power dissipation*, was just a pipe dream.

Then in 1961, Kunzler and his co-workers at Bell Telephone Laboratories discovered that specimens of Nb_3Sn wire, with the highest transition temperature yet found (18°K), could remain superconducting while carrying a current equivalent to an average current density of 10^5 amps cm^{-2}, in steady fields up to 88,000 oersted at 4·2°K (see Figure 4.4(b)). The critical field H_0 at $T = 0$, is believed to be in excess of 200 kOe.

This discovery changed a dream into reality, and within a year, a prototype superconducting solenoid of Nb_3Sn wire produced a field of 70 kOe.

Since then, many other high-field superconducting alloys and compounds have been found, most, like Nb_3Sn, being extremely brittle and therefore requiring sintering techniques to form into thin wires. The Ti Nb and Nb Zr alloys are ductile and can easily be drawn into wire, but have lower critical fields and currents (see Table 4.2). They are all non-ideal Type II superconductors, showing an incomplete Meissner effect. The extremely high critical fields are clearly a property of the mixed state divided into filaments so small that the distinction between alternate superconducting and normal phases is almost lost. The presence of considerable inhomogeneity, from compositional variation or strains introduced by cold working, appears necessary to stabilize the filamentary super-currents against the magnetic forces arising from neighbouring parallel filamentary super-currents, and from the solenoid field.

TABLE 4.2.

	T_c	H_N (at 0°K)	J_c (4·2°K)	at transverse field
	(°K)	(kOe)	(amp cm^{-2})	(kOe)
Nb_3Sn	18	~210	$>2 \times 10^5$	80
Ti Nb	9·7*	145*	2×10^4	80
Nb Zr	10·8*	130*	5×10^4	60
V_3 Ga	14·5	~270	10^5	80

Typical characteristics of some important Type II high-field superconductors.

*Highest value in solid solution range.

Commercial solenoids, capable of producing 60 kOe when cooled in liquid helium to 4·2°K, are now available, and small prototype solenoids have exceeded 100 kOe. Apart from requiring liquid helium, a super-conducting solenoid, or 'supermagnet', is simple to operate. Thus to energise a solenoid (see Figure 4.5) the 'superconducting switch' S is first 'opened' by heating a section of the superconducting wire shunt, connected across the solenoid current leads, until its temperature is above T_c. Switch K_1 is then closed and the current, from a 6v 20 amp supply (a 6v accumulator!), is increased to the desired value.

For the super-magnet to run in the persistent mode, S is 'closed' by switching off its heater, and allowing the shunt to cool below T_c. A persistent super-current now flows in the closed superconducting circuit, and the external supply may be disconnected by opening K_1.

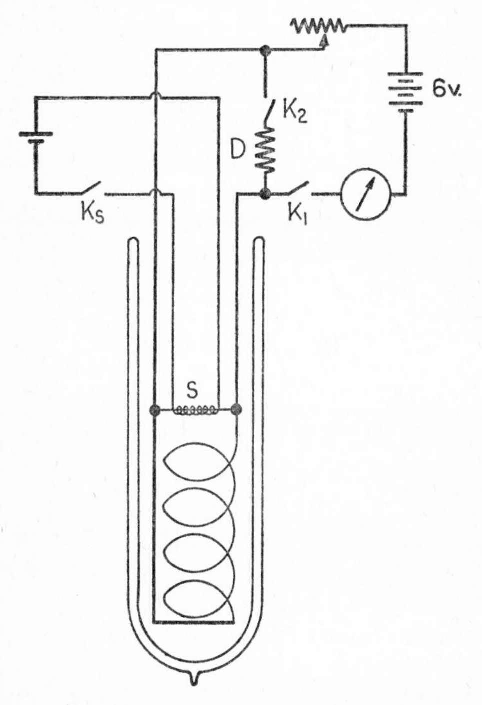

FIG. 4.5. A simple circuit for energizing a super-magnet.

The stored magnetic energy, $H^2/8\pi$ ergs cm^{-3}, is quite considerable, being 10 joules cm^{-3} at 50 kOe. If any section of the closed superconducting circuit goes normal, or 'quenches', for any reason, the appearance of a finite resistance will result in the stored magnetic energy being dissipated as heat within the solenoid. Furthermore, the rapid decay in the magnetic field will induce high voltages, and electrical breakdown through arcing may follow. A number of methods are being explored to slow down the decay of the magnetic field and to dump the magnetic energy outside the liquid helium cryostat at the onset of quenching, in order to avoid damage to the solenoid and explosive boil-off of liquid helium. One such method for protecting a super-magnet operating in the persistent mode is illustrated in Figure 4.5. An external 'dumping' resistance D is connected across the current leads by closing K_2. At the onset of quenching, a sensing device rapidly 'opens' the superconducting switch S, and a considerable part of the magnetic energy is dumped as heat in D, outside the liquid helium cryostat.

4.8. The Microscopic Theory of Superconductivity and the Energy Gap

The thermodynamic and electrodynamic properties do not, in themselves, tell us very much about the nature of superconductivity. The modern microscopic theory, first proposed by Bardeen, Cooper and Schrieffer (1951) and called the B.C.S. theory, goes a long way towards explaining, qualitatively at least, most of the properties of the superconducting state. *No* theory can yet explain *why* only certain metals, alloys and compounds become superconducting at low temperatures.

The B.C.S. theory predicts that a special type of attractive interaction between conduction electrons, opposing the normal electrostatic repulsion, can occur in the collective system of electrons and lattice taken together. The attraction takes place by the successive emission and absorption of short-lived 'virtual' phonons between pairs of electrons, and has the effect of pairing-off electrons with opposite spin and with approximately opposite wave vectors or linear momenta. The momenta of the electron pairs are distributed statistically about zero. The non-zero values of paired momenta, within the distribution, are equivalent to elementary super-currents.

Pairing is only effective for electrons close to the Fermi surface; i.e. those with energies lying within about kT_c of the Fermi energy W_F. The attraction, between an electron-pair, means that the total energy of a pair is lower, by a quantity 2ϵ, than the sum of the energies of the two unpaired electrons. At $T = 0$, all electrons close to the Fermi surface pair off, and an *energy gap* $2\epsilon_0$ ($= 3{\cdot}5kT_c$ from the theory) exists between the superconducting ground state and the normal excited states. At the same time, the density of states $N(W)$ are piled up on either side of the energy gap (see Figure 4.6(a)).

We must remember that the pairing off is a collective process involving all electrons near the Fermi-surface, together with the lattice through which they move. At finite temperatures, $0 < T < T_c$, some pairs are destroyed by thermal excitation across the energy gap into normal conduction states, and the strength of the collective pairing interaction 2ϵ is reduced. The energy gap becomes smaller and approaches zero as T approaches T_c (see Figure 4.6(b)). At T_c, the normal Fermi distribution is restored.

4.9. Evidence for the Superconducting Energy Gap

The transmission through a superconducting film, or reflection from the surface, of electromagnetic radiation in the far infra-red region (100–1000μ wavelength) suffers an abrupt drop in intensity when the phonon energy $h\nu$ exceeds the superconducting energy gap 2ϵ (Figure 4.6(c)). Increased absorption takes place when the phonon energy is sufficient to excite an electron pair across the gap into normal excited states. The absorption edge, as T approaches 0°K, indicates that $2\epsilon_0$ is about $3{\cdot}5\ kT_c$ in agreement with the B.C.S. theory.

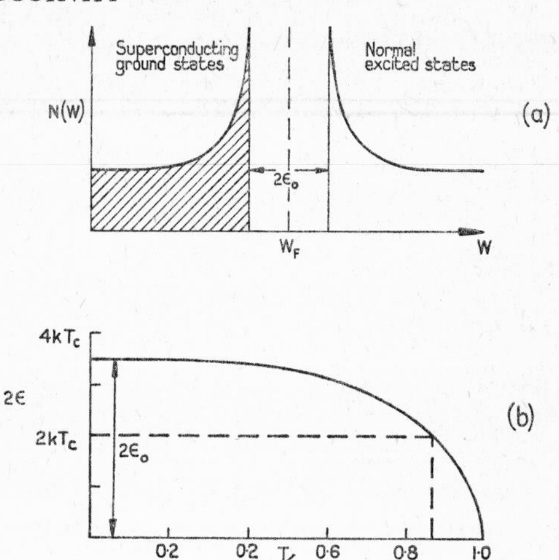

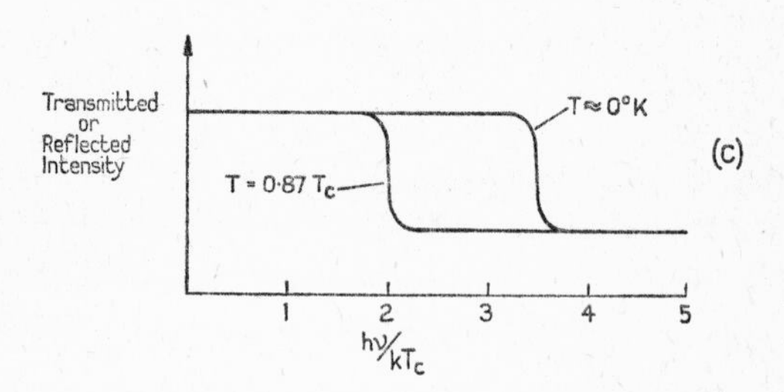

FIG. 4.6. The superconducting energy gap.

(a) Density of states $N(W)$ as a function of energy. At 0°K, all the shaded states are occupied. At higher temperatures, there is some excitation across the energy gap.

(b) Variation of the energy gap 2ϵ with T as predicted by the B.C.S. theory.

(c) Representative diagram of absorption against $h\nu/kT_c$ showing variation of absorption edge, and hence energy gap, with T.

4.10. The Energy Gap and Zero Resistance

The existence of an energy gap provides a simple, though incomplete, explanation for the phenomenon of zero resistance. Super electrons can only experience resistance to their motion by scattering processes in which the superconducting paired states are destroyed, i.e. when both electrons of a pair are excited across the energy gap into normal conduction states, requiring an extra energy of 2ϵ, of the order of $3{\cdot}5kT_c$.

At temperatures below T_c, the necessary thermal excitation energy is just not available for destroying a pair, and scattering cannot occur. The electrons in their superconducting paired states, therefore, pass freely through the lattice without being scattered, and hence with zero resistance to their motion.

4.11. Applications of Superconductivity

In addition to the use of high-field superconductors for high field, large volume super-magnets, described previously in Section 4.7, several other technological applications of superconductors may be mentioned.

1. The stored magnetic energy density in a 100 kOe solenoid is 40 joules cm^{-3}, comparable in magnitude with other electrical storage systems, and can readily be extracted as a short high-energy pulse.

2. With a critical current density in excess of 10^5 amps cm^{-2}, a high-field superconductor offers an attraction for high-power transmission with zero resistive loss. The cost of cooling a long superconducting transmission line with liquid helium might be economically worthwhile, one day.

3. Loss-free superconducting transformers are attractive, provided high-field superconductors are found with ideal reversible Type II behaviour, and no magnetic hysterisis.

4. At the other end of the energy scale, the low power superconducting switch or ‘cryotron’ has achieved importance as a

computer element. In its simplest form, a cryotron consists of a Ta wire, or gate, surrounded by a Nb coil, or control. At 4·2°K, both are superconducting, but Nb has a higher transition field than Ta. Therefore, a small current, through the Nb control, can produce sufficient field to quench the Ta gate wire into the normal resistive state, without quenching the Nb. Thus, a current passing through the gate wire may be controlled by a small one through the control, and the cryotron acts like a relay.

In this form, the self-inductance makes the switching time too slow for use as a computer element. Very much faster switching times ($\sim 10^{-9}$ sec) can be achieved with cryotrons made from evaporated thin films. The control and gate elements are in the form of crossed strips of film with different transition fields, separated by an insulating film. Thin film cryotrons complete with connecting circuits may be produced cheaply in large numbers, by suitable evaporation techniques, and offer possible advantages for cheap, reliable and compact computers.

5 Magnetic Phenomena

5.1. Introduction

Moving electric charges always have magnetic fields associated with them, and they experience magnetic forces in the presence of external magnetic fields. Atoms are made up of spinning and orbiting electric charges, and consequently, they are all affected in one way or another by an external magnetic field. The effect is described in terms of the magnetic dipole moment, or magnetisation per unit volume M, induced by a field H.

Three main types of behaviour take place:—

1. Diamagnetism, where M is small, in the opposite direction to H, and proportional in magnitude to H. All filled and unfilled shells of electrons in the atom are polarised diamagnetically, to some extent by an external field. Diamagnetic polarisation is analogous to the electric polarisation of an atom in an electric field, and is temperature independent. We shall, therefore, not discuss it.

2. Paramagnetism, where M is small, in the same direction as, and proportional in magnitude to H, for small fields.

3. Spontaneous magnetism, of which ferromagnetism is one example, where M is finite in zero field, is not proportional to H, and may approach values many orders of magnitude greater than those for paramagnetic substances. Other forms of spontaneous magnetism are antiferromagnetism and ferrimagnetism.

5.2. Paramagnetism

In the solid state, paramagnetism is a property of the permanent atomic magnetic moments arising from the inner unfilled electron shells of the transition elements. These include the iron group with an incomplete $3d$ shell, the palladium group (incomplete $4d$), the rare earth group (incomplete $4f$), the platinum group (incomplete $5d$) and the actinides (incomplete $5f$ shell).

Most experimental work has been done on the iron group and rare-earth group of elements, with their simpler chemistry. The three main methods of study depend on measurements from magnetic susceptibility, specific heat and paramagnetic spin resonance experiments. Let us first consider how the magnetic moment of an atom arises.

5.3. The Magnetic Moment of an Atom

Each electron in an atom is characterised by 4 quantum numbers (n, l, m, s). n is the principle quantum number and largely determines the electron energy. l and s determine the orbital angular momentum $\sqrt{l(l+1)} \cdot h/2\pi$ and spin angular momentum $\sqrt{s(s+1)} \cdot h/2\pi$ respectively, where l takes integer values from 0 to $(n-1)$, and s takes the values $+\frac{1}{2}$ and $-\frac{1}{2}$.

To illustrate how a magnetic dipole moment is associated with angular momentum, let us remember from electromagnetism that a current loop, area A, has a magnetic moment μ, perpendicular to the plane of the loop, given by

$$\mu = Ai$$

where i is the current flowing.

Consider an atomic electron moving in a circular orbit, radius a, with angular velocity $\boldsymbol{\omega}$ and orbital angular momentum

$$\mathbf{l} = ma^2\boldsymbol{\omega}$$

The moving electron is equivalent to an orbital current

$$i = -\frac{e}{2\pi} \times |\omega|$$

and has an associated magnetic moment, given by the vector equation,

$$\begin{aligned} \boldsymbol{\mu}_e &= -\pi a^2 \cdot \frac{e\boldsymbol{\omega}}{2\pi} \\ &= -\frac{e}{2m} \cdot \mathbf{l} \end{aligned} \qquad (5.1)$$

The magnetic moment associated with the spin angular momentum $\mathbf{s}$ of an electron cannot be derived so simply. The correct relation is

$$\boldsymbol{\mu}_s = -\frac{e}{2m} \cdot \mathbf{s} \times 2 \qquad (5.2)$$

where the extra factor of 2 (or more accurately 2·002319) arises from the relativistic nature of the spinning electron. The negative sign, in the two cases, arises from the negative electronic charge, and indicates that the associated magnetic moments are directed anti-parallel to the angular momentum vectors.

For light atoms, the orbital angular momenta of all the electrons in a sub-shell combine to give a resultant of magnitude $\sqrt{L(L+1)} \, . \, h/2\pi$, and similarly the resultant spin angular momentum is $\sqrt{S(S+1)} \, . \, h/2\pi$. These resultants then combine, vectorially, to form a total angular momentum of magnitude $\sqrt{J(J+1)} \, . \, h/2\pi$. L, S and J are orbital, spin and total quantum numbers, respectively, for the whole atom. From equations 5.1 and 5.2, the magnitude of the associated magnetic moments are:—

$$\mu_L = -\frac{e}{2m} \cdot \frac{h}{2\pi} \cdot \sqrt{L(L+1)} = -\beta\sqrt{L(L+1)}$$

and

$$\mu_s = -\frac{e}{m} \cdot \frac{h}{2\pi} \cdot \sqrt{S(S+1)} = -2\beta\sqrt{S(S+1)}$$

where $\beta \left(= \frac{eh}{4\pi m}\right)$ is the fundamental unit of atomic magnetic moment, called the Bohr magneton, with a value of

$$0{\cdot}93 \times 10^{-20} \text{ erg. gauss}^{-1}$$

Note that the magnetic moment per unit spin is twice that for each unit of orbital angular momentum. This leads to a complication when the two resultants combine vectorially to form the total angular momentum $\sqrt{J(J+1)} \,.\, h/2\pi$. The total magnetic moment μ_J is then expressed by

$$\mu_J = -g_J\beta\sqrt{J(J+1)} \tag{5.3}$$

where

$$g_J = \frac{3}{2} + \frac{S(S+1) - L(L+1)}{2J(J+1)}$$

and is called the Landé splitting factor.

For a filled shell, or sub-shell of electrons, L, S and J are zero, and no magnetic moment results. For an incompletely filled shell, J is non-zero and the whole atom possesses a magnetic moment given by equation 5.3. The free atom, with an unfilled valency shell, will generally possess a magnetic moment, and display paramagnetism in the free, unassociated, usually vapour, state. However, in a bound, associated, or chemically combined state, the magnetic moment of the valency shell disappears, on account of the nature of atomic bonds. (One notable exception is the paramagnetic oxygen molecule O_2). In the solid state, only atoms or ions with unfilled *inner shells* of electrons, i.e. the transition elements, are found to possess magnetic moments and display paramagnetism.

5.4. Effect of a Magnetic Field

In the presence of an external magnetic field, H, the total angular momentum vector $\sqrt{J(J+1)}h/2\pi$, and hence the magnetic moment μ_J of an atom, precess about the field direction, maintaining a constant angle with the field (Figure 5.1(a)). One

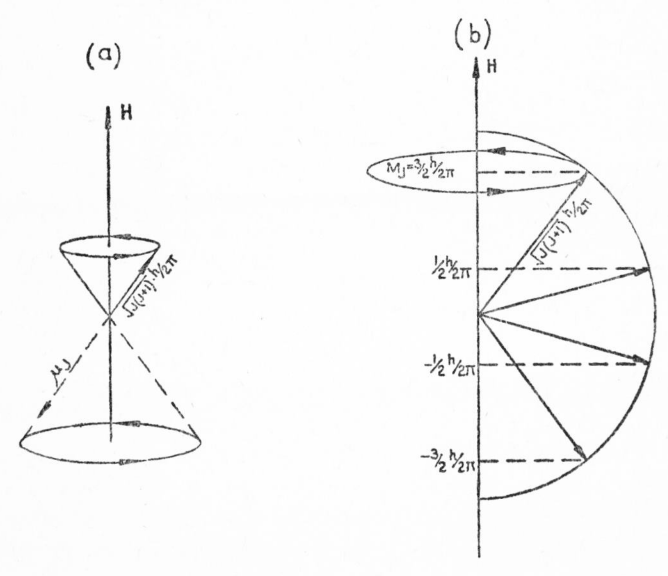

FIG. 5.1. (a) Precession of total angular momentum vector and associated atomic magnetic moment in a magnetic field.

(b) Quantization of total angular momentum in a magnetic field for the case $J = \frac{3}{2}$.

remarkable property of this precessing atomic magnet, is that the angle of precession, and hence the potential energy of the atomic dipole, is restricted to $2J + 1$ values—this is called *spatial quantisation* (Figure 5.1(b)). The quantisation is such that the projected value of the angular momentum along H has the values $M_J h/2\pi$, with M_J changing in integral steps between $-J$ and $+J$. Thus the steady, projected, components of the precessing magnetic moment in the direction of H is $-M_J g_J \beta$ and the potential energy (with $M_J = -J$ lowest) is:

$$W = M_J g_J \beta H$$

The energy splitting between adjacent magnetic levels is $g_J \beta H$, and is equivalent to a thermal energy of about 1°K when H is 10,000 oersted.

Now, provided the magnetic atoms are so far apart that their interaction energy is negligible in comparison with their thermal energy kT, the probability of an atom being in a state with energy W is proportional to $\exp(-W/kT)$. For a given value of M_J, this probability is proportional to $\exp(-M_J g \beta H/kT)$, (writing g for g_J) so that the fraction of atoms in this state is

$$\frac{\exp(-M_J g \beta H/kT)}{\sum \exp(-M_J g \beta H/kT)}$$

where the summation is taken over all M_J from $-J$ to $+J$.

The component of atomic magnetic moment parallel to H is $-M_J g \beta$, and the total magnetic moment M of one mole, containing N magnetic atoms, is

$$M = \frac{N \sum -M_J g \beta \exp(-M_J g \beta H/kT)}{\sum \exp(-M_J g \beta H/kT)}$$

where each summation is taken over all M_J.
This expression reduces to the form:

$$M = NgJ\beta\left[\frac{2J+1}{2J}\coth\left(\frac{2J+1}{2J}x\right) - \frac{1}{2J}\coth\left(\frac{x}{2J}\right)\right] \qquad (5.4)$$

where $x = Jg\beta H/kT$, and the expression in brackets is called the Brillouin function.

For $x \ll 1$, equation 5.4 reduces to the expression

$$M = Ng^2\beta^2\frac{J(J+1)}{3kT}H$$

In other words, the magnetic susceptibility χ, measured in small magnetic fields, is given by:

$$\chi = \frac{M}{H} = \frac{Ng^2\beta^2J(J+1)}{3kT} = \frac{C}{T} \quad (5.5)$$

where

$$C = \frac{Ng^2\beta^2J(J+1)}{3k} = \text{the Curie constant.}$$

This proportionality of χ with $1/T$ is known as Curie's law, and is obeyed by an 'ideal' paramagnetic substance, in which there is negligible interaction between magnetic atoms, and where

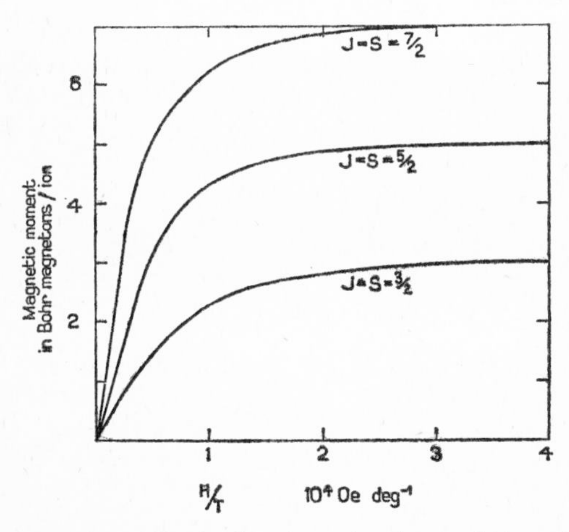

FIG. 5.2. The Brillouin function, showing variation of the average magnetic moment per atom with H/T, for $J = S = \frac{3}{2}$, $J = S = \frac{5}{2}$, and $J = S = \frac{7}{2}$,

the atomic magnetic levels are not split, in the absence of an external field, through interaction with crystalline electric fields in the solid state.

The behaviour of the Brillouin function is illustrated in Figure 5.2. For low values of x or H/T, the magnetisation varies linearly with H/T, i.e. in the region where Curie's law is obeyed. For larger values of x, where H/T is greater than 2×10^4 Oe deg^{-1}, the magnetisation approaches the saturation moment of $Ng\beta J$, when only the lowest magnetic level is thermally populated.

5.5. Paramagnetism below 1°K

The principle of attaining temperatures below 1°K by the adiabatic demagnetisation of a paramagnetic salt has been described in section 1.4.

Temperatures below 1°K are usually determined from the magnetic susceptibility χ of a paramagnetic salt, using Curie's law. Unfortunately, large deviations from the law occur at the lowest temperatures, and the measured 'magnetic temperature' $T^* = C/\chi$ cannot be assumed equal to the absolute temperature, T. Since in many cases, the deviations arise from the onset of spontaneous magnetism, it might be expected that a Curie-Weiss law (see section 5.10) would be sufficient to relate χ with absolute temperatures. In practice, this is not so, and published calibrations, of T^* against T for each salt, have to be used (Figure 5.3(a)). If these are not available, a rather laborious thermodynamic calibration of T^* in terms of T becomes necessary. Alternatively, the salt cerium magnesium nitrate, $Ce_2Mg_3(NO_3)_{12} \cdot 24\,H_2O$ (or C.M.N.), may be used for determining absolute temperatures. C.M.N. has a magnetic susceptibility obeying Curie's law very closely down to 0·007°K, and its magnetic temperature T^* is identical with T between 0·007°K and 4·2°K.

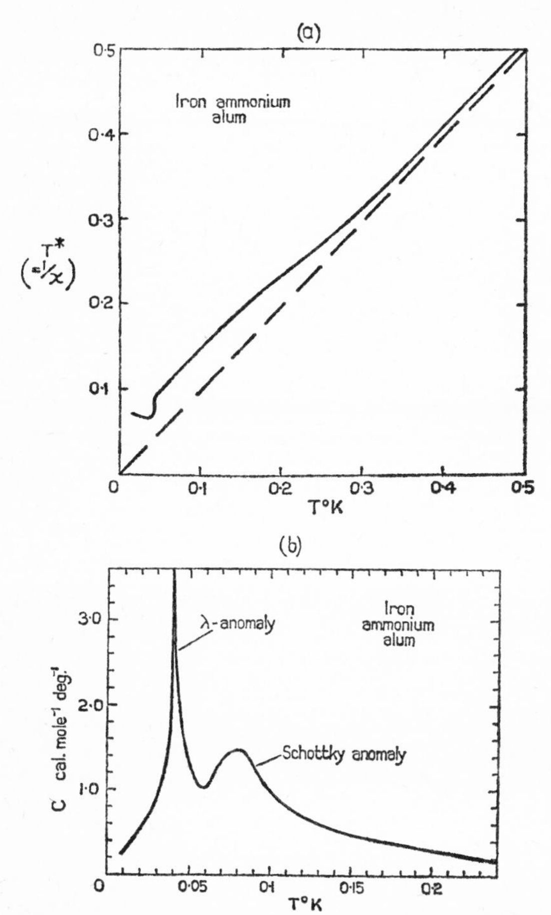

FIG. 5.3. (a) Variation of magnetic temperature T^* ($=1/\chi$) with absolute temperature T below 0·5°K for iron ammonium alum.

(b) Specific heat of iron ammonium alum below 0·2°K, showing λ-anomaly at 0·04°K and Schottky anomaly at 0·08°K (after Cooke and Kurti, 1949)

A good example of the behaviour of paramagnetic salts below 1°K is shown in the specific heat of ferric ammonium alum (Figure 5.3(b)). The λ-shaped co-operative anomaly indicates a transition to a spontaneously magnetised state (probably antiferromagnetic) at 0·04°K. The rounded Schottky anomaly, at 0·08°K, arises from the population of a magnetic level about 0·16°K above the ground state.

5.6. Nuclear Paramagnetism

All nuclei with odd numbers of protons, and/or neutrons, have non-zero values of nuclear spin angular momentum I (in half-integral units of $h/2\pi$, the same quantum unit of angular momentum as for electron spin). These include nuclei of all radioactive isotopes, and many isotopes of naturally occurring elements. Nuclei with even numbers of protons and neutrons have zero spin. The magnetic moment associated with non-zero spin of nuclei is about 2000 times smaller than the Bohr magneton β, being measured in terms of the nuclear magneton β_N, where

$$\beta_N = \frac{eh}{4\pi m_p} = \beta \times \frac{\text{electron mass}}{\text{proton mass}} = \frac{\beta}{1836}$$

Some values of nuclear spin, and magnetic moment, are given in Table 5.1. In a given magnetic field, the splitting of the nuclear magnetic levels is about 2000 times smaller than the corresponding splitting of electronic magnetic levels.

For the attainment of temperatures below 0·01°K, the adiabatic demagnetisation of nuclear magnetic spins has recently been achieved by Kurti, Simon and their co-workers at Oxford. For the appreciable removal of nuclear entropy, the initial magnetising condition, for H/T, must be greater, than that required for magnetic cooling with electron spins, by a factor of about 2000. The Oxford group have so far applied

TABLE 5.1.

Nucleus	Spin (h/2π)	Magnetic Moment (β_N)	
Proton $_1H^1$	$\frac{1}{2}$	+2·793	
Neutron $_0n^1$	$\frac{1}{2}$	−1·913	
$_{25}Mn^{55}$	$\frac{5}{2}$	+3·462	100% nat. abundance
$_{27}Co^{59}$	$\frac{7}{2}$	+4·64	100% nat. abundance
$_{27}Co^{60}$	5	+3·8	$\beta-\gamma$ radioactive
$_{28}Ni^{58}$	0	0	68% nat. abundance
$_{29}Cu^{63}$	$\frac{3}{2}$	+2·226	69% nat. abundance
$_{29}Cu^{65}$	$\frac{3}{2}$	+2·376	31% nat. abundance
$_{60}Nd^{143}$	$\frac{7}{2}$	−1·03	12% nat. abundance
$_{64}Gd^{155}$	$\frac{3}{2}$	−0·32	14·7% nat. abundance

Some values of nuclear spins and magnetic moments.

fields up to 40 kOe on copper nuclei maintained at an initial temperature of 0·01°K—a factor of about 400 greater. Although the entropy removal is small under these conditions, temperatures down to 10^{-6}°K have been attained.

The nuclear magnetic susceptibility χ_N will obey Curie's law,

$$\chi_N = \frac{N g_N^2 \beta_N^2 I(I+1)}{3k} \cdot \frac{1}{T}$$

$$= \frac{N \mu_N^2 \beta_N^2}{3k} \cdot \frac{(I+1)}{I} \cdot \frac{1}{T} \quad \text{since } g_N = \mu_N/I$$

The magnitude of χ_N is about 10^6 times smaller than electron magnetic susceptibilities, at the same temperature. Direct measurements of χ_N are possible, and have been used in the nuclear cooling experiments to determine temperatures between 10^{-6}°K and 10^{-3}°K.

The nuclear magnetic dipole-dipole interaction energy, μ_N^2/r^3, is some 10^6 times smaller than between electron magnetic moments, so that the lowest temperature attainable by nuclear cooling is around 10^{-7}°K, where spontaneous nuclear magnetism is expected to occur.

The influence of nuclear paramagnetism is quite considerable above the temperature range of nuclear cooling, between 0·01°K and 1°K. There is a strong internal magnetic interaction, between the nuclear moment and electron moment of the atom, in the paramagnetic or spontaneously magnetised state. This 'hyperfine' interaction is equivalent to an effective magnetic field seen by the nucleus, of magnitude between 100 and 1000 kOe, producing splittings between the nuclear magnetic levels, of the order of 0·01°K to 0·1°K. These splittings give rise to a Schottky type nuclear specific heat, with maxima ranging between 0·005°K and 0·05°K, and a $1/T^2$ tail extending above 1°K (e.g. Figure 5.4).

One consequence of the hyperfine splitting is that at 10^{-2}°K, the thermal population, of the nuclear magnetic levels, is concentrated in the lowest state. This corresponds to an orientation of the nuclei with respect to the electron magnetic moment. If the electron moments are aligned with a small external field—1 kOe is sufficient for a paramagnetic spin system, 5 kOe for a ferromagnetic system—then a spatial orientation, of all the nuclei in the magnetic atoms, will be achieved. If a γ-radioactive nucleus is included in the paramagnetic salt or ferromagnet, this nuclear orientation can be easily detected from the anisotropic distribution of γ-ray emission. The anisotropy, of γ-emission, with respect to the axis of orientation, is similar to the anisotropic radiation of radio-waves from a dipole aerial.

Nuclear orientation can be used either for the study of radioactive nuclear decay processes, or alternatively, since it is temperature dependent, as a secondary thermometer for the difficult temperature range below 1°K.

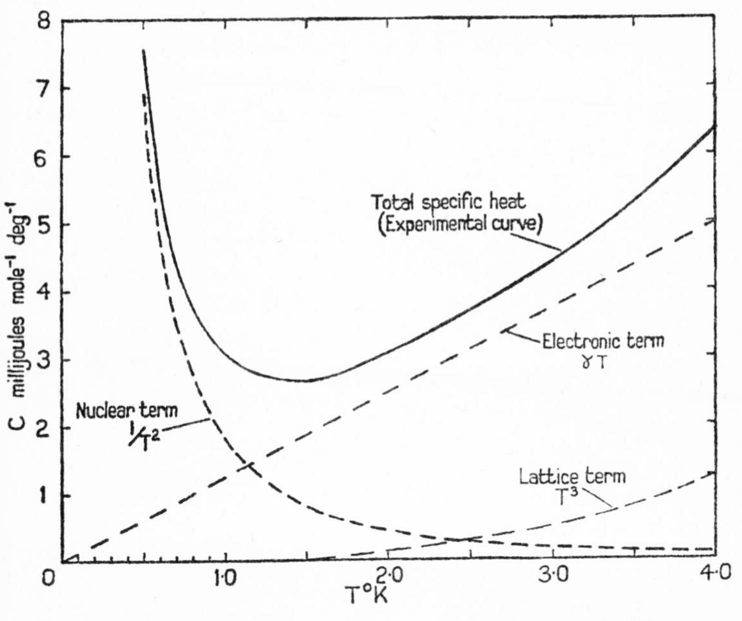

FIG. 5.4. Specific heat of MnNi between 0·5°K and 4°K showing $1/T^2$ specific heat tail due to the nuclear hyperfine interaction (from Scurlock and Wray, 1964)

α and β emission from oriented nuclei can also be studied, but the experimental techniques are more complex, since the α or β detector must be mounted inside the low temperature cryostat; whereas γ-ray intensity distributions are conveniently determined with counters mounted outside, and separate from, the cryostat. Study of β-emission, from oriented nuclei, provided the first demonstration of the non-conservation of parity in β-decay, from the experimental observation that β-particles are emitted preferentially from the 'tail or South pole' of Co^{60} nuclei.

5.7. Resonance Techniques

During the past 15 years, the detailed understanding of magnetic phenomena has been aided by the development of spectroscopic techniques, using electron spin resonance (E.S.R.) and nuclear magnetic resonance (N.M.R.).

Transitions between a magnetic energy level E_1 and a higher level E_2, separated by an energy gap $\Delta E = E_2 - E_1$, can be strongly stimulated by the oscillating magnetic component of a radio-frequency electromagnetic wave, when the frequency ν meets the resonance condition,

$$h\nu = \Delta E$$

Both up and down transitions, corresponding to absorption and emission, respectively, of electromagnetic radiation, are stimulated equally. A net absorption will occur if the number of atoms in the lower state, N_1, remains greater than the number, N_2, occupying the higher state. If the magnetic system remains in thermal equilibrium with the surrounding lattice, the absorption signal, proportional to $N_1 - N_2$, will depend on

$$[1 - \exp(-\Delta E/kT)],$$

or approximately $\Delta E/kT$, and increase with decreasing temperature. Thus, resonance experiments are favoured by working at low temperatures. Moreover, broadening of the magnetic levels, arising from thermal oscillations, is reduced at low temperatures, improving the sharpness, and hence resolution, of resonance 'lines'.

Generally, resonance is observed by including the paramagnetic sample in a resonant circuit, tuned to a fixed frequency, and varying the splitting of the magnetic levels with an external field. If the levels are unsplit in zero field, resonance absorption will take place when

$$h\nu = g_J \beta H$$

and be observed as a small dip in amplitude of oscillation of the resonant circuit (Figure 5.5). E.S.R. takes place at a frequency

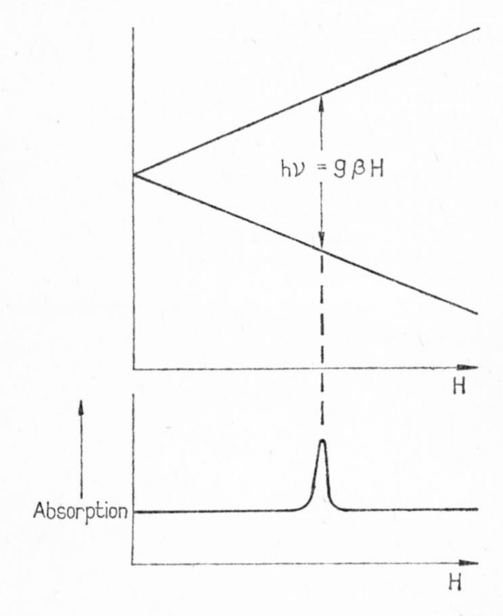

FIG. 5.5. Resonance absorption when $h\nu = g\beta H$

of the order of 10,000 Mc/s (3 cm wavelength, microwave region), in a field of 5 kOe; N.M.R. in the region of 5–30 Mc/s, in a similar field of 5 kOe.

The hyperfine splitting, between nuclear levels, gives rise to a hyperfine structure in E.S.R. spectra; until recently, most of our understanding of the hyperfine interaction had been gained from E.S.R. studies. However, it is now possible to observe the resonance absorption, from stimulated transitions between the split nuclear levels in a paramagnetic or ferromagnetic atom, in several other ways. These include variable frequency N.M.R. spectroscopy, Electron Nuclear Double Resonance

(E.N.D.O.R.), multiple pulse 'spin-echo' techniques, and resonant γ-ray absorption studies (Mössbauer effect), which are, today, leading to a far wider understanding of nuclear paramagnetism.

5.8. Spontaneous Magnetism

In general, all paramagnetic spin systems in the solid state will enter an ordered, or spontaneously magnetised, state below an ordering temperature characteristic of each system. This will take place when the interaction between neighbouring magnetic atoms, tending to line up their moments parallel or antiparallel to one another, is greater than the disorienting influence of the thermal energy.

There are three main types of spontaneous magnetism, illustrated for a one-dimensional chain of magnetic atoms in Figure 5.6(a). These are:—

1. Ferromagnetism, where the atomic dipoles become aligned parallel below a transition temperature called the Curie temperature T_c.

   The macroscopic magnetic behaviour of a ferromagnetic material is dominated by the formation of domain structure. Within each domain, of the order of 10^{-4} cm in size, all the atomic moments are aligned parallel.

2. Antiferromagnetism, the most common form of spontaneous magnetism, where neighbouring atomic moments become aligned antiparallel below a transition, or Néel, temperature T_N, so that the resultant magnetisation is zero. In a 3-dimensional crystal lattice, the number of antiparallel configurations is very large. This has been revealed by neutron diffraction experiments, which identify the magnetic lattice of a crystal in a way similar to the deter-

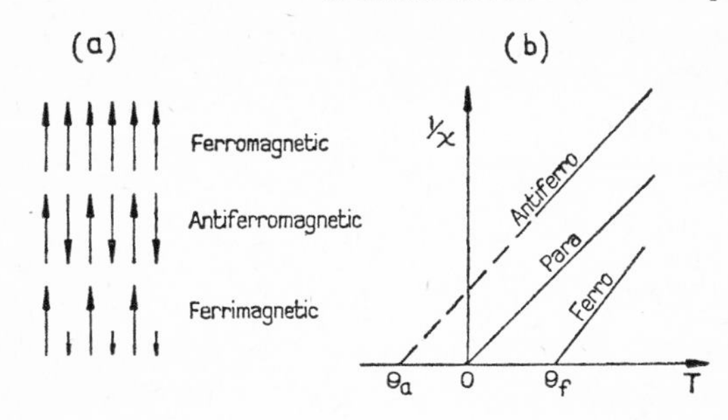

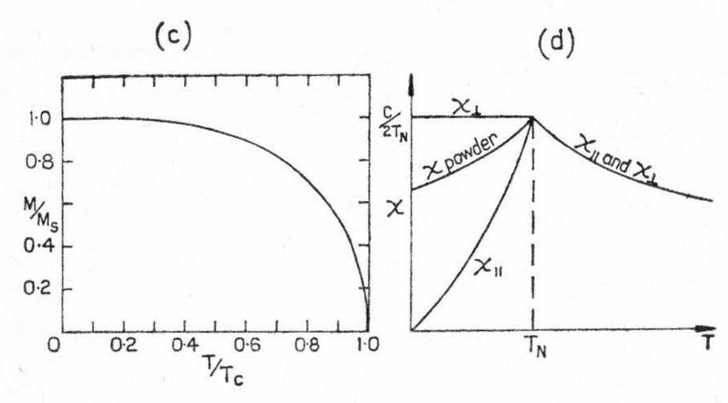

FIG. 5.6. (a) Ordering of magnetic moments in a 1-dimensional chain for ferromagnetism, antiferromagnetism and ferrimagnetism.

(b) Variation of $1/\chi$ with temperature for ferromagnetic, paramagnetic and antiferromagnetic solids.

(c) Typical variation of spontaneous magnetism M with temperature, for a ferromagnetic lattice, or an antiferromagnetic sub-lattice.

(d) Variation of $\chi_{\parallel}$ and $\chi_{\perp}$ for a single crystal, and χ_{powder} for a polycrystalline specimen, of an antiferromagnetic material.

mination of the atomic lattice by X-ray diffraction experiments.

3. Ferrimagnetism, a special case of antiferromagnetism, where two species of atomic moments, with different magnitudes, are aligned antiparallel, giving a resultant magnetisation comparable with that of ferromagnetic materials. Materials showing ferrimagnetism include garnets (e.g. yttrium iron garnet $Y_3Fe_5O_{12}$), and ferrites (with the general composition $X^{++}Fe_2{}^{+++}O_4$, where X is one, or a mixture, of divalent metals such as Mn, Co, Ni, Cu, Mg, Zn or Cd). The technical importance of ferrimagnets lies in their non-metallic nature and hence their application for high-frequency transformer cores, free from eddy-current losses.

5.9. The Ordering Interaction

There are two types of interaction between neighbouring atoms which give rise to spontaneous magnetism. They are illustrated, by considering the interaction between two neighbouring magnetic atoms, with spin angular momentum vectors $\mathbf{S}_1$ and $\mathbf{S}_2$, and magnetic moments $\boldsymbol{\mu}_1$ and $\boldsymbol{\mu}_2$, respectively. The two types are:—

1. Magnetic dipole-dipole interaction, tending to align the two moments, μ_1, and μ_2, antiparallel to one another in their state of lowest energy, and described by the interaction energy

$$W_{\text{mag}} = \frac{\boldsymbol{\mu}_1 \cdot \boldsymbol{\mu}_2}{r^3}$$

where r is the distance of separation.

The magnetic field at one atom ($\approx \mu/r^3$), due to a similar dipole moment of 1 Bohr magneton at a distance of 2×10^{-8} cm., is approximately 10^3 Oe. The corresponding difference in energy, between the lowest state of antiparallel alignment and the higher state of parallel alignment, is equivalent to a thermal energy of about 0·1°K. Thus, it can be expected that magnetic dipole-dipole interaction will not contribute to the occurrence of spontaneous magnetism above about 0·1°K.

2. Electrostatic exchange interaction, a quantum mechanical effect, which arises from overlapping of the electron charge distributions of the neighbouring magnetic atoms, and described in the form:

$$W_{ex} = -2J\mathbf{S}_1 \cdot \mathbf{S}_2$$

where J is a parameter called the exchange integral. The magnitude, and sign, of J is very sensitive to the degree of overlap of the electron charge distributions, and hence to the interatomic separation. When J is positive, we have ferromagnetic coupling, in which the spins align parallel in their state of lowest energy. When J is negative, we have antiferromagnetism with the adjacent spins aligning antiparallel. When J is zero, there is no exchange interaction and we can expect magnetic dipole-dipole interaction to produce spontaneous ordering at temperatures below 0·1°K.

The size of W_{ex} can be several orders of magnitude greater than W_{mag}. Thus exchange interaction, between magnetic atoms, is primarily responsible for the occurrence of spontaneous magnetism between 0·1°K and 1000°K.

The wide range of spontaneous magnetic ordering temperatures is illustrated in Table 5.2. Note how the ordering temperature decreases with increasing molecular weight, and hence separation of the magnetic atoms.

TABLE 5.2.

Magnetic Atom	Solid	Ordering Temperature (°K)	
Mn	α-Mn metal	100	antiferromagnetic
	Mn Cl_2 $4H_2O$	1·7	antiferromagnetic
	$(NH_4)_2SO_4MnSO_46H_2O$	0·14	ferromagnetic (?)
Fe	Fe metal	1043	ferromagnetic
	Fe Cl_2	24	antiferromagnetic
	$(NH_4)_2SO_4Fe_2(SO_4)_324H_2O$	0·04	antiferromagnetic
Cr	Cr metal	475	antiferromagnetic
	Cr Cl_2	70	antiferromagnetic
	$K_2SO_4Cr_2(SO_4)_324H_2O$	~0·01	ferromagnetic (?)

Spontaneous magnetic ordering temperatures of some solids.

5.10. The Weiss Field and Ferromagnetism

The most useful way of understanding ferromagnetism is to regard the exchange interaction as producing an internal magnetic field H_i, at each magnetic atom, which is proportional to the average magnetic moment of the neighbouring atoms, or the bulk magnetisation M. This Weiss field is a theoretical concept and should not be confused with the magnetic induction in a ferromagnetic material. For example, the theoretical value of H_i in ferromagnetic iron is about 10^7 gauss, very much larger than the magnetic induction, $\sim 10^4$ gauss.

On applying an external field H, Curie's law will still apply for small magnetisations, provided H is replaced by

$$H + H_i = H + \lambda M$$

We then have, $M = C(H + \lambda M)/T$
and the susceptibility

$$\chi = \frac{M}{H} = \frac{C}{T - C\lambda}$$

$$\text{or} \quad \chi = \frac{C}{T - \theta_f} \tag{5.6}$$

where $\theta_f = C\lambda$.

This is the Curie-Weiss law, and represents the behaviour of a paramagnetic substance at temperatures greater than θ_f, where θ_f is the Weiss constant (Figure 5.6(b)). For temperatures immediately below θ_f, the Weiss field starts to overcome the disorienting tendency of the thermal energy of the magnetic spins, and the spins tend to line up parallel without the aid of an external field. This increases M, and hence H_i becomes, in turn, even larger. In this way, the spontaneous magnetisation M increases rapidly over a small drop in temperature, just below the transition or Curie temperature T_c. M continues to rise, with decreasing temperature, to a saturation value M_s at absolute zero, in a way characteristic of most ferromagnetic materials (Figure 5.6(c)).

At the same time, a λ specific heat anomaly, typical of a co-operative order-disorder transition, occurs. The position of the specific heat peak is usually used to define T_c, as distinct from θ_f derived from susceptibility data.

In general, the specific heat anomaly does not fall to zero until some temperature well above T_c is reached. This indicates that, although long range magnetic order disappears at T_c, short range ordering persists to a higher temperature. The latter is borne out by neutron diffraction experiments above T_c and supported by the fact that θ_f is slightly greater than T_c.

5.11. Antiferromagnetism

In antiferromagnetism, where the spins align antiparallel, we can picture the spins as arranged on two interpenetrating sub-lattices; each sub-lattice bearing spins all parallel, but with the spins on one sub-lattice being anti-parallel to those on the other. A similar internal field treatment may be used, in which an atom, on sub-lattice A, experiences a field from its neighbours on sub-lattice B, proportional to the B sub-lattice magnetisation M_B.

Then, in an applied field H, the total field H_A on the A sub-lattice atom is

$$H_A = H - H_{iB} = H - \lambda M_B$$

and similarly

$$H_B = H - \lambda M_A$$

Above the Néel temperature T_N, we can calculate the total magnetic susceptibility, using H_A and H_B, and remembering that half the atoms are on each sub-lattice.

Then

$$M_A = \tfrac{1}{2}C(H - \lambda M_B)/T \tag{5.7}$$

and

$$M_B = \tfrac{1}{2}C(H - \lambda M_A)/T \tag{5.8}$$

The total magnetisation per unit volume,

$$\begin{aligned} M &= M_A + M_B \\ &= \tfrac{1}{2}C[2H - \lambda(M_A + M_B)]/T \end{aligned}$$

and the susceptibility

$$\chi = \frac{M}{H} = \frac{C}{T + \theta_a} \tag{5.9}$$

where

$$\theta_a = \frac{\lambda C}{2} \tag{5.10}$$

Thus above the Néel temperature, an antiferromagnet obeys a Curie-Weiss law, with a Weiss constant θ_a *opposite* in sign to the ferromagnetic Weiss constant θ_f (Figure 5.6(b)).

The transition temperature T_N is that below which each sub-lattice possesses a spontaneous magnetisation in the absence of an external field. From equation (5.10), $\lambda = 2\theta_a/C$. Substituting λ in equations (5.7) and (5.8), and assuming that Curie's Law is obeyed at T_N, we have

$$M_A = -M_B \cdot \frac{\theta_a}{T_N} \text{ and } M_B = -M_A \frac{\theta_a}{T_N}$$

for which the non-trivial solution is $\theta_a = T_N$.
Experimental values of θ_a/T_N lie in the range 1·5 to 5, indicating that this simple model is insufficient. Much better agreement is obtained, by taking into account the additional interaction between next nearest neighbours on the same sub-lattice.

A co-operative λ-specific heat anomaly is also associated with an antiferromagnetic transition, the peak co-inciding with the Néel temperature T_N (Figure 5.7). In practice, the existence of a λ-specific heat anomaly is one of the first indications of an anti-ferromagnetic transition. This is because the magnetic properties of the antiferromagnetic state are not startlingly different from the paramagnetic state—unlike ferromagnetism.

The variation of each sub-lattice magnetisation with temperature below T_N is very similar to that of a ferromagnet as in Figure 5.6(c). However, in zero external field, the two sub-lattice magnetisations are equal and opposite, and the resultant

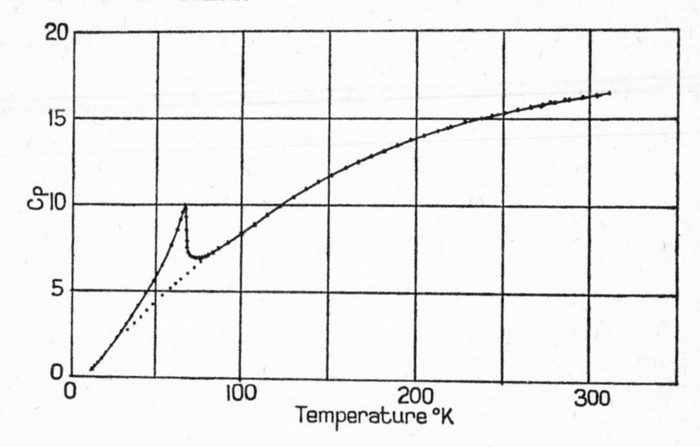

FIG. 5.7. Specific heat of MnF showing specific heat anomaly associated with antiferromagnetic transition at 70°K (from Stout and Adams, 1942, by kind permission of the American Chemical Society)

magnetisation is zero. The variation of magnetic susceptibility with temperature is indicated in Figure 5.6(d) for single crystal, and polycrystalline, antiferromagnetic specimens.

6 Mechanical Properties at Low Temperatures

6.1. Introduction

The study and classification of the mechanical properties of materials at low temperatures have only recently been stimulated on a wide scale, by the increasing commercial application of low temperatures. Quite naturally, the interest has been almost entirely confined to high strength metals and alloys.

The fundamental interaction between atoms in a solid lattice can be expressed in terms of the elastic constants, relating the elastic forces between the atoms. However, the resistance to deformation under load, or the shear strength of a metal, is two or three orders of magnitude less than that expected theoretically from the elastic constants. This inherent weakness of metals arises from the presence of particular types of imperfection in the lattice. The behaviour of these imperfections, called dislocations, therefore, largely determines the mechanical properties of metals.

The variation of the elastic constants with temperature is very small, the constants generally increasing by a few per cent between 300°K and 4·2°K. Very much larger variations of the other mechanical properties with temperature are observed, and these can be associated with temperature dependent effects arising from the lattice imperfections. To a good approximation the slight increase in elastic constants at low temperatures can be neglected.

6.2. Dislocations

The tensile strength, fatigue and creep properties are all largely determined by the motion and mutual interaction of dislocations in the crystal lattice. Very briefly, a dislocation is a special type of one-dimensional lattice imperfection which either forms a closed loop, or extends to the crystal boundaries. When a dislocation is made to glide on a plane through a crystal under the influence of an applied stress, atoms on either side of the plane move relative to one another by a distance **b**, of the order of an atomic spacing. The direction of **b** relative to the dislocation line characterises the dislocation; if **b** is perpendicular, to the dislocation line, we have an *edge* dislocation; if **b** is parallel to the dislocation line, we have a *screw* dislocation (Figure 6.1(a) and (b)).

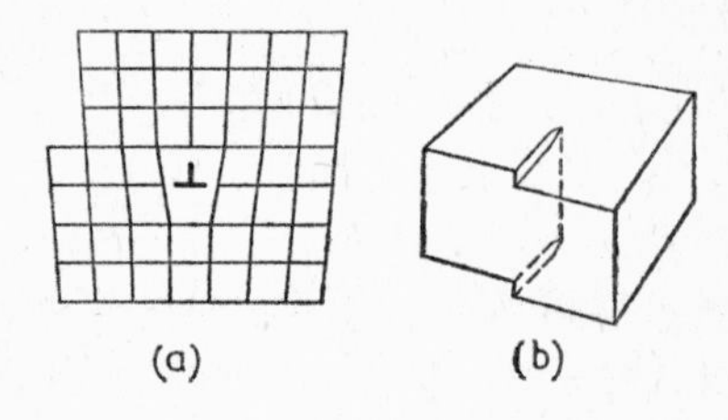

FIG. 6.1. (a) An edge dislocation. The dislocation line is perpendicular to plane of paper.
(b) A screw dislocation. The dislocation line is along the vertical broken line.

Slip, of one portion of a crystal with respect to the other, is equivalent to the successive passage of dislocations along the slip plane from one side of the crystal to the other, each dislocation producing a slip **b**. In a perfect lattice, a dislocation can move under the influence of a very small stress, of the order of 10^6 dynes cm^{-2}, about 10^{-5} times the shear modulus. Thus, a pure metal single crystal is extremely soft, and deforms easily when a stress is applied. In normal metals, met with in practice however, there are always obstacles tending to hold up the movement of dislocations and so reducing the tendency to deform. The most important of these are:—

1. Impurity atoms, which are attracted to dislocations, forming a cloud around them and restricting their motion.
2. Neighbouring dislocations, whose stress fields act either by attraction or repulsion to restrict motion.
3. 'forest' dislocations which intersect the slip plane of the gliding dislocation.

The presence of these obstacles means that a greater stress is required to move dislocations right through the crystal and produce deformation. At high temperatures, the thermal activation energy ($\sim kT$ per atom), associated with the thermal vibrations of the crystal lattice, helps the applied stress to move the dislocations past the obstacles. Hence the external stress required to produce a given deformation at high temperatures is less than that required at low temperatures. For this reason, most metals become stronger and harder as the temperature is reduced.

The density of dislocations is usually specified by the number of dislocation lines intersecting unit area of crystal, and ranges from 10^2–10^3 cm^{-2} in the most pure crystals to as many as 10^{12} cm^{-2} in heavily deformed crystals. The latter figure is equivalent to one dislocation per square of side 100 atoms.

6.3. Tensile Properties

6.3(a) *General Outline.* All materials strain elastically under increasing stress up to the yield point, beyond which deformation takes place, in the sense that the deformation remains permanent after reducing the stress to zero. The performance of a material beyond the yield point largely determines the working load and usefulness of that material.

In general, two types of behaviour occur at low temperatures:

1. plastic deformation, down to the lowest temperatures, occurring in face-centred cubic (f.c.c.) and hexagonal close packed (h.c.p.) metals, e.g. Al, Cu and stainless steels.
2. brittle fracture, occurring below a temperature T_B, with little plastic deformation; confined mainly to body centred cubic (b.c.c.) metals, e.g. iron and low alloy steels, but not the alkali metals.

6.3(b). *Plastic deformation.* Typical stress-strain curves for metal single crystals (e.g. copper), which remain ductile down to 4·2°, are shown in Figure 6.2 (a). The deformation behaviour

(a)

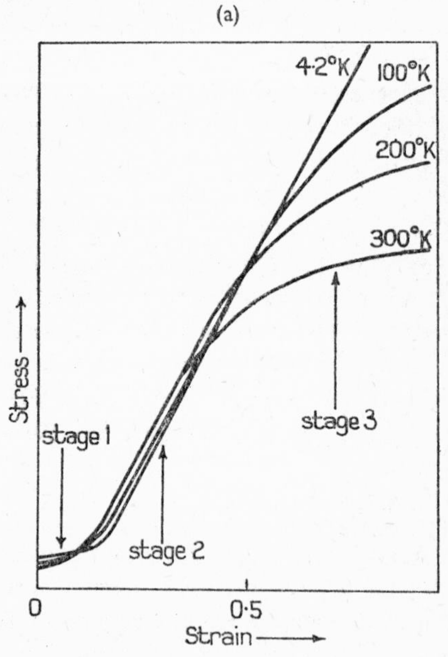

FIG. 6.2. (a) Idealized stress-strain curves for a f.c.c. metal at various temperatures, showing the three stages of plastic deformation (from Rosenberg, 1963)

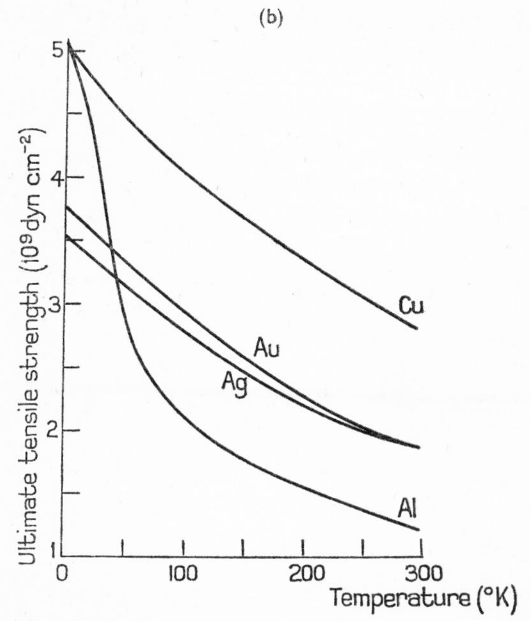

FIG. 6.2. (b) Increase in tensile strength of Cu, Au, Ag and Al at low temperatures (from McCammon and Rosenberg, 1957)

changes with increasing stress and may be divided into three stages. After the yield stress is exceeded (this may increase by up to 50% at 4·2°K), Stage 1 is entered, where slip occurs along one or more slip planes. This corresponds to the movement of dislocations along easy-glide planes, with little work hardening.

Stage 2 is the region of rapid work hardening, where plastic deformation is proportional to stress. In this region, it is believed that the dislocations rapidly multiply and pile up behind one or other type of obstruction. In polycrystalline samples, the small size and random orientation of the crystallites, or grains, prevent Stage 1 deformation, and Stage 2 is observed as the initial behaviour.

The onset of Stage 3 is strongly temperature dependent. It is believed to arise when the sum of stress energy and thermal activation energy is sufficient to enable the piled up groups of dislocations, created in Stage 2, to move past their respective obstructions. Once the stress is sufficiently high to move the dislocation past one obstacle, the dislocations will continue to pass other similar obstacles, and so deformation becomes progressively larger as the stress is further increased. Since at low temperatures, the thermal activation energy is less, a greater stress is necessary before Stage 3 commences.

The tensile stress, at which fracture occurs, is approximately determined when the decrease in cross-section is no longer compensated by the increase in work-hardening of the metal. This condition will start to apply after the onset of Stage 3, when fracture usually occurs. The strong temperature dependence, of the onset of Stage 3, means that the tensile strength is increased at lower temperatures. Figure 6.2(b) shows this general increase of the tensile strength for polycrystalline samples. An outstanding example is aluminium, where the tensile strength at 4·2°K is about four times the strength at 300°K.

6.3(c). *Brittle Fracture*. In some metals, the yield stress increases much more steeply with decreasing temperature than that observed in metals showing plastic deformation. Although these metals are ductile at room temperature, they become brittle below a certain temperature T_B. Once the yield stress is exceeded below T_B, little plastic deformation takes place and brittle fracture occurs.

Most of these metals have b.c.c. structure, and include iron, and manganese or carbon steels. Fortunately, stainless steels with high nickel content do not exhibit brittle fracture, probably because the structure is not b.c.c.

The difference in behaviour, of the yield strength of f.c.c. and b.c.c. lattices, arises from the different ways in which edge and screw dislocations interact with impurity atoms in the two lattices. In the f.c.c. structure, an impurity atom produces a spherical distortion of the lattice, which can only be relieved by an edge dislocation. A cloud of impurity atoms is, then, attracted *only* by edge dislocations, leaving screw dislocations free to move under lower stresses. Thus the yield strength of f.c.c. metals is little affected by temperature.

In the b.c.c. structure, however, an impurity atom causes non-spherical distortion of the lattice, which can be relieved by both screw and edge dislocations. Consequently, *both* types of dislocation become anchored by impurity clouds. Before slip can occur, considerable stress must first be applied to pull the dislocation from its cloud. This stress will be aided by thermal activation at high temperatures, but not at low, and so the yield stress increases considerably at low temperatures (in a similar way to the onset of Stage 3 plastic deformation).

Since brittle fracture takes place in single crystal, as well as polycrystalline, samples, it is believed to arise from the creation of micro-cracks within the crystal, as well as at grain boundaries. These micro-cracks are common to all crystal lattice structures and probably arise from the combination of dislocations. Once these cracks are formed, the applied stress at the head of the cracks is large enough to make them spread, unless plastic deformation takes place to relieve the stress field around the crack. As the temperature is decreased, and the yield stress rises, a point is reached when plastic deformation no longer takes place, the cracks spread, and brittle fracture occurs.

6.4. Creep at Low Temperatures

When a metal is subjected to a constant stress, it deforms continuously with time. This phenomenon is called creep, and is sensitive to temperature. At low temperatures around 90°K, the creep strain ϵ, for a constant stress, varies according to the relation:

$$\epsilon = ET \log gt \tag{6.1}$$

where E and g are constants, T is the absolute temperature and t is the time from when the stress was first applied.

The temperature dependence can be explained by assuming that dislocations are held up at obstacles, which they can only pass with the aid of thermal activation. As the temperature rises, the probability of thermal activation increases, and creep continues more rapidly.

The few measurements made at liquid helium temperatures show that creep becomes temperature independent, and is very much larger than that predicted by equation 6.1. This additional creep at helium temperatures may be explained by a quantum mechanical tunnelling of dislocations past their obstructions.

6.5. Fatigue at Low Temperatures

Fatigue occurs when an alternating stress is applied to a ductile metal. After a certain number of cycles, fatigue fracture takes place, even though the peak stress applied may be considerably less than that required to produce yield or fracture in a static test.

Although fatigue is a complex process, it appears to arise from the initial creation of micro-cracks within the grains. This is followed by the growth and combination of these micro-cracks, until the effective cross-sectional area is sufficiently reduced for fracture to occur.

The mechanism of formation and growth of the micro-cracks is not understood very well. The growth is accelerated by thermal activation processes, as shown by the experimental observations that the fatigue lifetime, under a given alternating stress, increases very considerably as the temperature is reduced

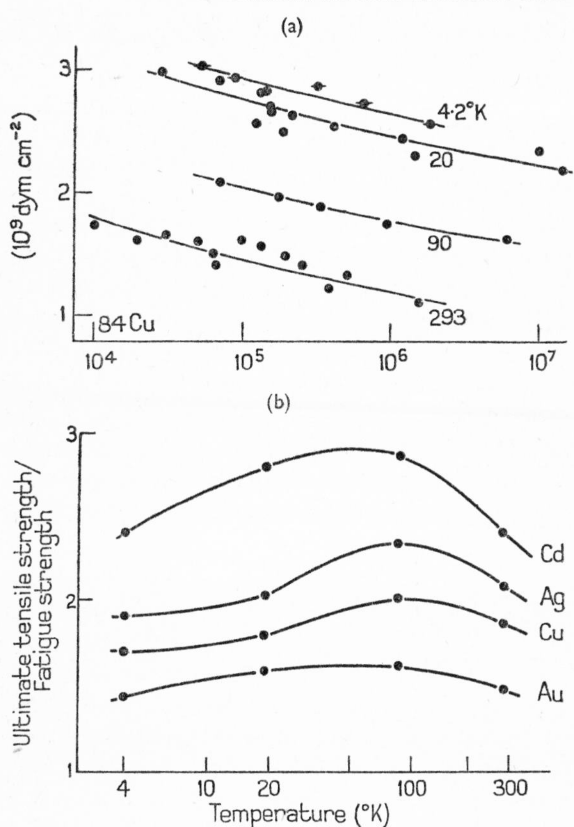

FIG. 6.3. (a) Fatigue in copper at low temperatures showing that as the temperature is reduced, there is (i) a 100–1000 fold increase in fatigue life at constant stress, and (ii) an increase in fatigue stress for failure in given number of cycles. Each experimental point is a plot of peak alternating stress against number of cycles to fatigue fracture.

(b) Variation of the ratio, ultimate tensile strength/fatigue strength for Cd, Ag, Cu and Au.

(from McCammon and Rosenberg, 1957)

—by a factor of 10^2 or 10^3 between 300°K and 4·2°K (Figure 6.3(a)).

It is interesting to consider the ratio, at different temperatures, of the tensile strength to the peak stress required to produce fatigue fracture in a given number of cycles (Figure 6.3(b)). The ratio is surprisingly constant, and suggests that similar dislocation mechanisms are responsible for the increase in both tensile and fatigue strengths at low temperatures.

Examination, of the fractured surfaces after fatigue, indicates that fatigue fracture takes place in the same way at 300°K and 4·2°K. Although diffusion and corrosion processes may contribute towards micro-crack formation and the general fatigue behaviour at 300°K, these processes are unlikely to occur at 4·2°K. It may, therefore, be concluded that micro-crack mechanisms, and hence fatigue failure, are associated primarily with the interaction or combination of dislocations, under the influence of alternating stress.

Suggestions for further reading

For general reading

BATES, L. F. (1961) *Modern Magnetism* (Cambridge University Press).

BLEANEY, B. I. and BLEANEY, B. (1957) *Electricity and Magnetism* (Clarendon Press, Oxford).

DEKKER, A. J. (1957) *Solid State Physics* (Macmillan).

HOARE, F. E., JACKSON, L. C. and KURTI, N. (1961) *Experimental Cryophysics* (Butterworth).

KITTEL, C. (1956) *Introduction to Solid State Physics* (Wiley).

ROSENBERG, H. M. (1963) *Low Temperature Solid State Physics* (Clarendon Press, Oxford).

SCHOENBERG, D. (1952) *Superconductivity* (Cambridge University Press).

VANCE, R. W. (1963) *Cryogenic Technology* (Wiley).

WHITE, G. K. (1959) *Experimental Techniques in Low Temperature Physics* (Clarendon Press, Oxford).

WILKS, J. (1961) *The Third Law of Thermodynamics* (Oxford University Press).

For particular topics

Chapter 1.

SIMON, F. E., KURTI, N., ALLEN, J. F. and MENDELSSOHN, K. (1952) *Low Temperature Physics; Four Lectures* (Pergamon).

Chapter 2.

PARKINSON, D. H. (1958) 'The specific heat of metals at low temperatures'. *Repts. on Progr. in Phys.* **21**, 226.

Chapter 3.

BERMAN, R. (1953) 'The thermal conductivity of dielectric solids at low temperatures'. *Adv. in Phys.* **2**, 103.

ROSENBERG, H. M. (1955) 'The thermal conductivity of metals at low temperatures'. *Phil. Trans.* **247**, 441.

Chapter 4.

BERLINCOURT, T. G. (1963) 'High magnetic fields by means of superconductors'. *Brit. J. Appl. Phys.* **14**, 749.

MENDELSSOHN, K. (1963) 'Patterns of superconductivity'. *Cryogenics* **3**, 129.

Chapter 5.

AMBLER, E. and HUDSON, R. P. (1955) 'Magnetic cooling'. *Repts. on Progr. in Phys.* **18**, 251.

KURTI, N. (1958) 'Nuclear orientation and nuclear cooling'. *Physics Today* **11**, 19.

Chapter 6.

COTTRELL, A. H. (1953). *Dislocations and Plastic Flow in Crystals.* (Clarendon Press, Oxford).

ROSENBERG, H. M. (1958). 'The properties of metals at low temperatures'. *Progress in Metal Physics* **7**, 339.

Index